Mathematical Analysis of Random Phenomena

Proceedings of the International Conference

Mathematical Analysis of Random Phenomena

Proceedings of the International Conference

Hammamet, Tunisia 12 – 17 September 2005

Editors

Ana Bela Cruzeiro
Grupo de Física–Matemática & Universidade Técnica de Lisboa, Portugal

Habib Ouerdiane
University of Tunis El Manar, Tunisia

Nobuaki Obata
Tohoku University, Japan

World Scientific

NEW JERSEY • LONDON • SINGAPORE • BEIJING • SHANGHAI • HONG KONG • TAIPEI • CHENNAI

Published by

World Scientific Publishing Co. Pte. Ltd.

5 Toh Tuck Link, Singapore 596224

USA office: 27 Warren Street, Suite 401-402, Hackensack, NJ 07601

UK office: 57 Shelton Street, Covent Garden, London WC2H 9HE

British Library Cataloguing-in-Publication Data
A catalogue record for this book is available from the British Library.

MATHEMATICAL ANALYSIS OF RANDOM PHENOMENA
Proceedings of the International Conference

ISBN-13 978-981-270-603-4
ISBN-10 981-270-603-8

Printed in Singapore by World Scientific Printers (S) Pte Ltd

Preface

This volume contains research articles resulting of the "International Conference on Mathematical Analysis of Random Phenomena" that took place in Hammamet, Tunisia, from 12 to 17 September 2005 and was Coorganized by the Portuguese Mathematical Society (SPM) and the Tunisian Mathematical Society (SMT).

This meeting was devoted to the exposition of recent developments in the mathematical analysis of random phenomena: stochastic analysis and its applications, mathematical physics, infinite dimensional analysis, probability theory and their interactions.

One can read in this volume eighteen articles on the following topics: stochastic analysis and infinite dimensional analysis, white noise analysis, Malliavin calculus and applications, mathematical finance, Poisson analysis, hydrodynamics, statistical mechanics, and probability in quantum physics.

We are grateful to all Tunisian and Portuguese institutions which have brought to the scientific organizing committee their moral and financial supports. These are, in particular,

· Ministère Tunisien de l´Enseignement Supérieur,
· Ministère de la Recherche Scientifique, de la Technologie et du Développement des Compétences, Tunisia GRICES,
· Université de Tunis El Manar,
· University of Lisbon, Grupo de Física Matemática,
· University of Madeira, Centro de Ciências Matemáticas,
· Société Mathématique de Tunisie,
· Sociedade Portuguesa de Matemática,
· Faculté des Sciences de Tunis.

We wish also to thank the authors for their contribution to a book of high quality, accessible to a large scientific public, as well as the colleagues who helped us with their anonymous and careful referee work.

The Editors,

Ana Bela Cruzeiro, Nobuaki Obata, Habib Ouerdiane
Lisbon / Sendai / Tunis, July 12, 2006

Contents

GEOMETRY AND INTEGRATION BY PARTS ON $H \setminus \mathrm{Diff}(S^1)$

HÉLÈNE AIRAULT (*INSSET, Saint-Quentin / LAMFA, Amiens*)

We study various tensor fields on the Lie algebra $\mathrm{diff}(S^1)$ and we give their expressions in the trigonometrical basis. We define a bounded operator Ψ on $\mathrm{diff}(S^1)$ modulo su(1, 1). With Ψ, we obtain an integration by parts formula on $H\setminus\mathrm{Diff}(S^1)$.

0. Introduction

In [1], the Levi-Civita connection on $H \setminus \mathrm{Diff}(S^1)$, the quotient space $\mathrm{Diff}(S^1)$ modulo the homographic transformations has been explicited in real and complex coordinates and the existence of the parallel transport on $H \setminus \mathrm{Diff}(S^1)$ has been established. With an extension to the infinite dimensional case of the Fang-Malliavin structure equations (see [12]), integration by parts formulae have been obtained on $H \setminus \mathrm{Diff}(S^1)$. In the first part, we deepen the study of the geometry of tensor fields on $\mathrm{diff}(S^1)$ as started in [6], [7], [8], [15], [16], [4], [1], [14]. Both manifolds $H \setminus \mathrm{Diff}(S^1)$ and $\mathrm{Diff}(S^1)/\mathrm{Rot}(S^1)$ have a structure of Kählerian manifold. In [6], [7], [8], [15], [16], [4], [1], [14], the Ricci tensor has been proved to be a diagonal operator in the trigonometrical basis. It is a multiple of the metric tensor and these manifolds are Einstein manifolds. The bracket on the Lie algebra $\mathrm{diff}(S^1)$ is defined by

$$[u, v] = uv' - u'v \quad \text{for} \quad u \in \mathrm{diff}(S^1), \quad v \in \mathrm{diff}(S^1) \tag{1}$$

and the Hilbert transform is linear and it is given on trigonometric functions by

$$J \cos k\theta = \sin k\theta \quad \text{and} \quad J \sin k\theta = -\cos k\theta \quad \text{for } k \geq 1, \quad J\,1 = 0. \tag{2}$$

In [1], it has been proved in particular that with the Levi-Civita tensor field Γ, the operators $\Gamma(\cos p\theta)^2 + \Gamma(\sin p\theta)^2$ are diagonal operators in the trigonometrical basis. In this work, in relation with the Levi-Civita tensor field on $H \setminus \mathrm{Diff}(S^1)$, we study tensor fields $\Gamma(u) : \mathrm{diff}(S^1) \to \mathrm{diff}(S^1)$ which (i) commute with J, (ii) are torsionless, (iii) for $p \geq 1$, the operators $\Gamma(\cos p\theta)^2 + \Gamma(\sin p\theta)^2$ are diagonal operators in the trigonometrical basis

$(e_p)_{p\geq 1} = \{\cos k\theta, \sin k\theta\}_{k\geq 1}$; we resume these three conditions as

$$\begin{aligned} &\text{(i)} \quad \Gamma(u)v - \Gamma(v)u = [u,v] \ \text{(torsionless condition)} \\ &\text{(ii)} \quad \Gamma(u)Jv = J\Gamma(u)v \end{aligned} \qquad \text{for } u, v \in \mathrm{diff}(S^1), \tag{3}$$

$$\text{(iii)} \quad \begin{cases} [\Gamma(\cos p\theta)^2 + \Gamma(\sin p\theta)^2] \sin k\theta = \lambda_{p,k} \sin k\theta, \\ [\Gamma(\cos p\theta)^2 + \Gamma(\sin p\theta)^2] \cos k\theta = \lambda_{p,k} \cos k\theta. \end{cases} \tag{4}$$

Let

$$\alpha(k) = ak^3 + bk \quad \text{where } a > 0; \tag{5}$$

then

$$\begin{aligned} (m+j)\alpha(m-j) + (j-2m)\alpha(j) + (2j-m)\alpha(m) = 0, \\ (2j+m)\alpha(m) = (m-j)\alpha(m+j) + (2m+j)\alpha(j). \end{aligned} \tag{6}$$

We shall take $\alpha(k) = k^3 - k$. It satisfies $(k+2)\alpha(k) = (k-1)\alpha(k+1)$. Conversely, it is remarked in section 3 of the present work that the condition (6) on $\alpha(k)$ implies that $\alpha(k)$ is of the form (5) up to a constant term. On $\mathrm{diff}(S^1)$, let $(\ |\ )$ be a pseudo metric defined by the conditions, for $k \geq 1$, $(\cos k\theta \mid \cos k\theta) = (\sin k\theta \mid \sin k\theta) = \alpha(k)$, and $\{\cos k\theta, \sin k\theta\}_{k\geq 0}$ are orthogonal vectors on $\mathrm{diff}(S^1)$. The interest of this metric is due first to its close relationship and well adaptedness to the trigonometrical basis, secondly to its remarkable properties. In particular the second fundamental two-form is closed. In [3], it has been proved that the metric $\alpha(k) = k^3 - k$ is the unique one (up to the multiplication by a constant) such that $Ad(h)$ is unitary for any homographic transformation h. In the following, we extend the construction of diagonal operators associated to the Levi-Civita tensor field, (see [1], section 2) to more general tensor fields Γ. We study the tensor fields Γ satisfying (4) and such that

$$\begin{aligned} \Gamma(\cos p\theta) \sin k\theta = \beta_p(k) \cos(p+k)\theta + \gamma_p(k)\, 1_{p\geq k+1} \cos(p-k)\theta \\ + \mu_p(k)\, 1_{k\geq p+1} \cos(k-p)\theta + a_k\, \delta_k^p \cos k\theta \end{aligned} \tag{7}$$

where δ_k^p is the Kronecker symbol. We obtain that, for $p \geq 1$, the operator $\Gamma(\cos p\theta)^2 + \Gamma(\sin p\theta)^2$ is diagonal in the trigonometric basis $\{\cos k\theta, \sin k\theta\}_{k\geq 1}$ if and only if $\gamma_p(k)\, 1_{p\geq k+1} = 0$. In the same way, the operator

$$\Phi_p \ :\ u \to [\Gamma(\cos p\theta)u, \cos p\theta] + [\Gamma(\sin p\theta)u, \sin p\theta] \tag{8}$$

is diagonal in the trigonometric basis $\{\cos k\theta, \sin k\theta\}_{k\geq 1}$ if and only if $\gamma_p(k)\, 1_{p\geq k+1} = 0$. In that case, $\mathbf{Z}_p$ defined by $\mathbf{Z}_p = \Phi_p + \Gamma(\cos p\theta)^2 +$

$\Gamma(\sin p\theta)^2$ is diagonal and $\mathbf{Z}_p \sin k\theta = \lambda_p^Z(k) \sin k\theta$, $\mathbf{Z}_p \cos k\theta = \lambda_p^Z(k) \cos k\theta$ with

$$\lambda_p^Z(k) = -1_{k \geq p+1}\,(p+k)\left[\frac{1}{2}(2p-k) + \beta_p(k-p)\right] - k\,a_k \delta_k^p. \tag{9}$$

If we assume that $\beta_p(k) = \frac{(2p+k)\alpha(k)}{2\alpha(p+k)}$, this value of $\beta_p(k)$ gives the Levi-Civita connection, then the trace

$$2\sum_{p\geq 0} \lambda_p^Z(k) = -\frac{13}{6} - \frac{2k(a_k - 1)}{\alpha(k)} \tag{10}$$

is finite. In particular, when Γ is the Levi-Civita tensor field, we study the operator Ψ defined by

$$\Psi = \sum_j [\Gamma(\epsilon_j) + \phi(\epsilon_j)]\,\Gamma(\epsilon_j) \tag{11}$$

where for h, $u \in \mathrm{diff}(S^1)$,

$$\phi(\epsilon_j)h = [\epsilon_j, h]. \tag{12}$$

The operator Ψ is bounded on $\mathrm{diff}(S^1)$. For more general Γ satisfying (3), we obtain more bounded operators on $\mathrm{diff}(S^1)$ with finite trace as (10).

The second part is stochastic analysis. With Ψ and an adaptation of Fang's integration by parts on loop groups [11], we establish the following integration by parts formula for the Levi-Civita connection on $H\setminus\mathrm{Diff}(S^1)$. Consider the canonical Brownian motion $x(t)$ on $\mathrm{diff}(S^1)$, see [17], [2], [13]. Let $\Omega(\bullet)$ be the stochastic parallel transport of the frames above $H \setminus \mathrm{Diff}(S^1)$. See [1] for the existence of $\Omega(\bullet)$. We have the SDE

$$dz(t) = \sum_\alpha \Omega_\alpha(t) * dx_\alpha(t)\,; \qquad d\Omega_\alpha^k(t) + \Gamma(\Omega_\alpha(t), *dz(t), \epsilon_k) = 0 \quad \forall\alpha, \forall k\,. \tag{13}$$

A continuous map $t \to \gamma(t)$ from $[0, +\infty[$ to $\mathcal{H} \setminus \mathrm{Diff}(S^1)$ is denoted $\gamma(\bullet)$. We call X the Wiener space of such continuous maps. For $h \in V$ and $\mu \in R$,

$$(\exp(\mu\, h))(\theta) = \theta + \mu h(\theta) + \cdots \tag{14}$$

We have

$$\begin{aligned} &\exp(-\mu\Omega(t)h)\,o\,d\,\exp(\mu\Omega(t)h) \\ &\quad = (I - \mu\Omega(t)h)\,o\,\mu\,d\Omega(t)h\, + \text{ terms in } \mu^j, \quad j \geq 2 \\ &\quad = \mu\, d\Omega(t)\,h + \text{terms in } \mu^j, \quad j \geq 2. \end{aligned} \tag{15}$$

In the same way,

$$d\,\exp(\mu\Omega(t)h)\,o\,\exp(-\mu\Omega(t)h) = \mu d\Omega(t)h + \text{terms in } \mu^j, \quad j \geq 2. \tag{16}$$

We consider the process

$$\gamma_\mu(t) = \exp(\mu\Omega(t)h)\,o\,\gamma(t). \tag{17}$$

For $F : X \to R$, we put

$$D_h^\Gamma F(\gamma(\bullet)) = \frac{d}{d\mu}\Big|_{\mu=0} F(\exp(\mu\Omega(\bullet)h) o \gamma(\bullet)). \tag{18}$$

Denoting $(\,|\,)$ the metric on V, then

$$E[(D_h^\Gamma F)(\gamma(\bullet))] = E\left[F(\gamma(\bullet)) \int_0^1 (\Psi\,\Omega(t)h \mid o\,dx(t))\right]. \tag{19}$$

Part I. Geometry on trigonometric functions

1. Bracket, metric and structure constants

Let δ_i^j be the Kronecker symbol, with the bracket (1), we have

$$\begin{cases} 2[\cos k\theta, \cos p\theta] = \sum_{q\geq 1}[(k-p)\delta_q^{k+p} + (k+p)\delta_q^{k-p} - (k+p)\delta_q^{p-k}]\sin q\theta, \\ 2[\sin k\theta, \sin p\theta] = \sum_{q\geq 1}[(p-k)\delta_q^{p+k} + (k+p)\delta_q^{k-p} - (k+p)\delta_q^{p-k}]\sin q\theta, \\ 2[\cos k\theta, \sin p\theta] = \sum_{q\geq 1}[(p-k)\delta_q^{k+p} + (k+p)\delta_q^{k-p} + (k+p)\delta_q^{p-k}]\cos q\theta \\ \qquad\qquad + 2k\delta_k^p. \end{cases} \tag{20}$$

In complex coordinates, for $m, n \in Z$, $[e^{im\theta}, e^{in\theta}] = i\,(n-m)e^{i(m+n)\theta}$. From (20),

$$\begin{cases} (i) \quad [\sin k\theta, \sin p\theta] - [\cos k\theta, \cos p\theta] = (p-k)\sin(k+p)\theta, \\ (ii) \quad [\cos k\theta, \sin p\theta] + [\sin k\theta, \cos p\theta] = (p-k)\cos(k+p)\theta. \end{cases} \tag{21}$$

We take $\alpha(k) = k^3 - k$. For $j \geq 2$, we put

$$c_j = \frac{\cos j\theta}{\sqrt{\alpha(j)}}, \quad s_j = \frac{\sin j\theta}{\sqrt{\alpha(j)}}. \tag{22}$$

For $j \geq 2,\ k \geq 2$,

$$\begin{cases} 2[c_j, c_k] = \sum_{p\geq 2} K^p_{jk} s_p + \left[\frac{\delta_j^{k+1}(2k+1)}{\sqrt{\alpha(k)}\sqrt{\alpha(k+1)}} - \frac{\delta_j^{k-1}(2k-1)}{\sqrt{\alpha(k)}\sqrt{\alpha(k-1)}} \right] \sin\theta, \\ 2[s_j, s_k] = \sum_{p\geq 2} S^p_{jk} s_p + \left[\frac{\delta_j^{k+1}(2k+1)}{\sqrt{\alpha(k)}\sqrt{\alpha(k+1)}} - \frac{\delta_j^{k-1}(2k-1)}{\sqrt{\alpha(k)}\sqrt{\alpha(k-1)}} \right] \sin\theta, \\ 2[s_j, c_k] = \sum_{p\geq 2} (SC)^p_{jk} c_p \\ \qquad\qquad - \left[\frac{\delta_j^{k+1}(2k+1)}{\sqrt{\alpha(k)}\sqrt{\alpha(k+1)}} + \frac{\delta_j^{k-1}(2k-1)}{\sqrt{\alpha(k)}\sqrt{\alpha(k-1)}} \right] \cos\theta + 2k\delta_k^j, \end{cases}$$

where K^p_{jk}, S^p_{jk} are antisymmetric in (j,k),

$$\begin{aligned} K^p_{jk} &= \frac{(j-k)\sqrt{\alpha(p)}}{\sqrt{\alpha(j)}\sqrt{\alpha(k)}}\delta_p^{j+k} + \frac{(2k+p)\sqrt{\alpha(p)}}{\sqrt{\alpha(j)}\sqrt{\alpha(k)}}\delta_j^{k+p} - \frac{(2j+p)\sqrt{\alpha(p)}}{\sqrt{\alpha(j)}\sqrt{\alpha(k)}}\delta_k^{j+p}, \\ S^p_{jk} &= \frac{(k-j)\sqrt{\alpha(p)}}{\sqrt{\alpha(j)}\sqrt{\alpha(k)}}\delta_p^{j+k} + \frac{(2k+p)\sqrt{\alpha(p)}}{\sqrt{\alpha(j)}\sqrt{\alpha(k)}}\delta_j^{k+p} - \frac{(2j+p)\sqrt{\alpha(p)}}{\sqrt{\alpha(j)}\sqrt{\alpha(k)}}\delta_k^{j+p}, \\ (SC)^p_{jk} &= \frac{(k-j)\sqrt{\alpha(p)}}{\sqrt{\alpha(j)}\sqrt{\alpha(k)}}\delta_p^{j+k} - \frac{(2k+p)\sqrt{\alpha(p)}}{\sqrt{\alpha(j)}\sqrt{\alpha(k)}}\delta_j^{k+p} - \frac{(2j+p)\sqrt{\alpha(p)}}{\sqrt{\alpha(j)}\sqrt{\alpha(k)}}\delta_k^{j+p}. \end{aligned}$$

2. Tensor fields on diff(S^1), their expressions in the trigonometrical basis

The Hilbert transform J possesses the Nijenhuis property with respect to the Lie bracket (1). For $u,\ v \in V$,

$$[Ju, Jv] - [u,v] = J([u, Jv] + [Ju, v]). \tag{23}$$

We define the following tensor fields, for $u,\ v \in \mathrm{diff}(S^1)$, see [4],

$$\begin{cases} E(u,v) = [Ju, Jv] - [u,v], \\ F(u,v) = [u, Jv] + [Ju, v], \end{cases} \qquad \begin{cases} G(u,v) = [Ju, Jv] + [u,v], \\ H(u,v) = [u, Jv] - [Ju, v]. \end{cases} \tag{24}$$

We remark that E, F, G are antisymmetric in u, v, whereas H is symmetric in u, v. On the other hand, (23) is the same as

$$E(u,v) - J\,F(u,v) = 0. \tag{25}$$

We obtain

$$E(Ju, Jv) = -E(u, v) \quad \text{and} \quad \begin{aligned} H(u,v) &= G(u, Jv) = -G(Ju, v), \\ [u,v] &= -\frac{1}{2}(E(u,v) - G(u,v)), \\ J[Ju,v] &= \frac{1}{2}(E(u,v) - JH(u,v)). \end{aligned} \tag{26}$$

We put

$$A(u,v) = J[u,v] - [Ju,v] \quad \text{and} \quad B(u,v) = J[u,v] - [u,Jv]. \tag{27}$$

Since $J^2 = -1$, we have when $u \neq v$,

$$\begin{aligned} A(u,v) &= JA(Ju,v) \\ A(u,Jv) + A(Ju,v) &= 0 \end{aligned} \quad \text{and} \quad B(u,v) = JB(u,Jv). \tag{28}$$

A and B are neither symmetric nor antisymmetric in u, v, the decomposition in symmetric and antisymmetric part is given by

$$\begin{cases} 2A(u,v) = JG(u,v) + H(u,v), \\ 2B(u,v) = JG(u,v) - H(u,v). \end{cases} \tag{29}$$

We express these different tensor fields in the trigonometrical basis, we have

$$\begin{aligned} F(\cos k\theta, \cos p\theta) &= (p-k)\cos(p+k)\theta, \\ F(\cos k\theta, \sin p\theta) &= (p-k)\sin(p+k)\theta, \end{aligned} \quad \text{and} \quad \begin{aligned} F(Ju, Jv) &= -F(u,v), \\ JF(u,v) &= F(u, Jv). \end{aligned} \tag{30}$$

For $j \geq 1$, $k \geq 1$,

$$\begin{aligned} H(\sin j\theta, \cos k\theta) &= (k+j)\sin(j-k)\theta, \\ H(\cos j\theta, \cos k\theta) &= (k+j)\cos(j-k)\theta, \end{aligned} \quad \text{and} \quad H(Ju, Jv) = H(u,v). \tag{31}$$

The symmetric tensor

$$Q(u,v) = \frac{1}{2}\, J\, H(u,v) = \frac{1}{2} J([u, Jv] + [v, Ju]) \tag{32}$$

has been found by [14]. For $j \geq 1$, $k \geq 1$,

$$\begin{aligned} 2Q(\sin j\theta, \cos k\theta) &= -(1_{j\geq k} - 1_{k\geq j})(k+j)\cos(k-j)\theta \\ 2Q(\cos j\theta, \cos k\theta) &= (1_{j\geq k} - 1_{k\geq j})(k+j)\sin(j-k)\theta \\ &\text{and } Q(Ju, Jv) = Q(u,v). \end{aligned} \tag{33}$$

In the same way,

$$\begin{cases} G(\sin j\theta, \cos k\theta) = -(k+j)\cos(k-j)\theta \\ G(\cos j\theta, \cos k\theta) = (k+j)\sin(j-k)\theta \end{cases} \quad \text{and} \quad G(Ju, Jv) = G(u,v) \tag{34}$$

and

$$P(u,v) = \frac{1}{2}JG(u,v) = Q(Ju,v) = \frac{1}{2}J([Ju,Jv]+[u,v]) \tag{35}$$

satisfies

$$\begin{aligned} &2P(\sin j\theta, \cos k\theta) = (1_{j\geq k} - 1_{k\geq j})(k+j)\sin(k-j)\theta, \\ &2P(\cos j\theta, \cos k\theta) = -(1_{j\geq k} - 1_{k\geq j})(k+j)\cos(j-k)\theta, \\ &P(Ju, Jv) = P(u,v). \end{aligned} \tag{36}$$

For A, B,

$$\begin{cases} A(\sin j\theta, \cos k\theta) = -1_{k\geq j}(k+j)\sin(k-j)\theta \\ A(\cos j\theta, \cos k\theta) = 1_{k\geq j}(k+j)\cos(k-j)\theta \end{cases} \quad \text{and } A(Ju, Jv) = A(u,v), \tag{37}$$

$$\begin{cases} B(\sin j\theta, \cos k\theta) = -1_{j\geq k}(k+j)\sin(j-k)\theta \\ B(\cos j\theta, \cos k\theta) = -1_{j\geq k}(k+j)\cos(k-j)\theta \end{cases} \quad \text{and } B(Ju, Jv) = B(u,v). \tag{38}$$

Let

$$D(u,v) = \frac{1}{2}([Ju,Jv]+[u,v]) = \frac{1}{2}G(u,v); \tag{39}$$

then

$$\begin{cases} Q(u,v) + D(u,v) = [u,v] + J[u,Jv] = -JB(u,v), \\ Q(u,v) - D(u,v) = -[u,v] - J[Ju,v] = JA(u,v). \end{cases} \tag{40}$$

On the other hand, let

$$\{u,v\} = \frac{1}{2}([u,Jv]+[Ju,v]); \tag{41}$$

then

$$\begin{aligned} 2\{\cos k\theta, \cos p\theta\} &= (p-k)\cos(p+k)\theta, \\ 2\{\sin k\theta, \sin p\theta\} &= (k-p)\cos(p+k)\theta, \\ 2\{\cos k\theta, \sin p\theta\} &= (p-k)\sin(p+k)\theta. \end{aligned} \tag{42}$$

3. The fundamental two-form and the metric

The identities (6) give the closure condition of the symplectic form on $\operatorname{diff}(S^1)$. It is the same as

$$(m-n)\alpha(p)+(p-n)\alpha(m)+(m-p)\alpha(n)=0 \quad \text{with} \quad m+n+p=0 \tag{43}$$

or equivalently

$$\det\begin{pmatrix}1 & x & \alpha(x) \\ 1 & y & \alpha(y) \\ 1 & -(x+y) & \alpha(-(x+y))\end{pmatrix}=0. \tag{44}$$

We look for solutions of (44) and assume that α has no singularity at zero. The function $\alpha(x) = \lambda x + \mu$ is a solution of (44). We may assume that $\alpha(0) = 0$ and $\alpha'(0) = 0$. With $y = 0$ in (44), we obtain that α is an odd function. In (44), we take the derivative with respect to y, and we put $y = 0$ and $\alpha'(0) = 0$; we obtain $-3\alpha(x) + x\alpha'(x) = 0$, thus $\alpha(x) = bx^3$.

With (6), we obtain (see [4]),

$$([u,v] \mid Jw)+([w,u] \mid Jv)+([v,w] \mid Ju)=0 \tag{45}$$

and for the fundamental two-form Φ

$$\Phi(u,v)=(u \mid Jv) \tag{46}$$

it gives

$$\Phi([u,v],w)+\Phi([w,u],v)+\Phi([v,w],u)=0. \tag{47}$$

4. The Levi-Civita connection on $H \setminus \operatorname{Diff}(S^1)$

Let $u \in V$, $v \in \operatorname{diff}(S^1)$, $w \in V$, we define the Levi-Civita tensor field (see [1]) $\Gamma_1(v)u$ with

$$2(\Gamma_1(v)u \mid w)=([w,v] \mid u)+([w,u] \mid v)-([u,v] \mid w)) \tag{48}$$

and we put

$$2(\Lambda_1(v)u \mid w)=-2(\Gamma_1(u)v \mid w)=([v,w] \mid u)+([u,w] \mid v)+([v,u] \mid w) \tag{48'}$$

then

$$\Gamma_1(u)v+[v,u]=\Gamma_1(v)u. \tag{49}$$

We have

$$\Gamma_1(v)Ju = J\Gamma_1(v)u \quad \text{and} \quad (\Gamma_1(v)u \mid w) = -(u \mid \Gamma_1(v)w). \tag{50}$$

Both Γ_1 and Λ_1 are torsionless. The expression of Γ_1 in the trigonometrical basis has been given in [1]. If $v = 1$, $v = \cos\theta$, $v = \sin\theta$, then $2(\Gamma_1(v)u \mid w) = ([w,v] \mid u) - ([u,v] \mid w)$, thus $2(\Gamma_1(1)u \mid w) = -(w' \mid u) + (u' \mid w) = 2(u' \mid w)$.

$$\Gamma_1(1)u = u', \tag{51}$$

$$\begin{cases} 2\Gamma_1(\cos\theta)\cos k\theta = -1_{k\geq 3} \times (k+1)\sin(k-1)\theta - (k-1)\sin(k+1)\theta, \\ 2\Gamma_1(\cos\theta)\sin k\theta = 1_{k\geq 3} \times (k+1)\cos(k-1)\theta + (k-1)\cos(k+1)\theta, \end{cases} \tag{52}$$

$$\begin{cases} 2\Gamma_1(\sin\theta)\sin k\theta = -1_{k\geq 3} \times (k+1)\sin(k-1)\theta + (k-1)\sin(k+1)\theta, \\ 2\Gamma_1(\sin\theta)\cos k\theta = -1_{k\geq 3} \times (k+1)\cos(k-1)\theta + (k-1)\cos(k+1)\theta, \end{cases} \tag{53}$$

and, for $p \geq 2$,

$$\begin{cases} 2\Gamma_1(\cos p\theta)(\sin k\theta) \\ \quad = (p+k)1_{k\geq p+2}\cos(k-p)\theta + \dfrac{(2p+k)\alpha(k)}{\alpha(p+k)}\cos(p+k)\theta, \\ 2\Gamma_1(\cos p\theta)(\cos k\theta) \\ \quad = -(p+k)1_{k\geq p+2}\sin(k-p)\theta - \dfrac{(2p+k)\alpha(k)}{\alpha(p+k)}\sin(p+k)\theta, \end{cases} \tag{54}$$

$$\begin{cases} 2\Gamma_1(\sin p\theta)(\sin k\theta) \\ \quad = -(p+k)1_{k\geq p+2}\sin(k-p)\theta + \dfrac{(2p+k)\alpha(k)}{\alpha(p+k)}\sin(p+k)\theta, \\ 2\Gamma_1(\sin p\theta)(\cos k\theta) \\ \quad = -(p+k)1_{k\geq p+2}\cos(k-p)\theta + \dfrac{(2p+k)\alpha(k)}{\alpha(p+k)}\cos(p+k)\theta. \end{cases} \tag{55}$$

We put (see [14])

$$\Lambda_2(v)u = \Lambda_1(v)u - Q(u,v) \quad \text{and} \quad \Gamma_2(v)u = -\Lambda_2(u)v. \tag{56}$$

Thus

$$\Gamma_2(v)u = \Gamma_1(v)u + Q(u,v); \tag{57}$$

we have

$$\begin{aligned} 2(\Lambda_2(v)u \mid w) &= ([w,Jv] \mid Ju) + ([w,Ju] \mid Jv) - ([u,v] \mid w), \\ 2(\Gamma_2(v)u \mid w) &= -2(\Lambda_2(u)v \mid w) \\ &= ([Jv,w] \mid Ju) + ([Ju,w] \mid Jv) - ([u,v] \mid w). \end{aligned} \tag{58}$$

We obtain

$$\Lambda_2(v)Ju = J\Lambda_2(v)u \tag{59}$$

and both Λ_2 and Γ_2 are torsionless. For $p \geq 2$,

$$\begin{cases} 2\Lambda_2(\cos p\theta)(\sin k\theta) \\ \quad = (p+k)1_{k\geq p+2}\cos(k-p)\theta - \dfrac{(2k+p)\alpha(p)}{\alpha(p+k)}\cos(p+k)\theta, \\ 2\Lambda_2(\cos p\theta)(\cos k\theta) \\ \quad = -(p+k)1_{k\geq p+2}\sin(k-p)\theta + \dfrac{(2k+p)\alpha(p)}{\alpha(p+k)}\sin(p+k)\theta, \end{cases} \tag{60}$$

$$\begin{cases} 2\Lambda_2(\sin p\theta)(\sin k\theta) \\ \quad = -(p+k)1_{k\geq p+2}\sin(k-p)\theta - \dfrac{(2k+p)\alpha(p)}{\alpha(p+k)}\sin(p+k)\theta, \\ 2\Lambda_2(\sin p\theta)(\cos k\theta) \\ \quad = -(p+k)1_{k\geq p+2}\cos(k-p)\theta - \dfrac{(2k+p)\alpha(p)}{\alpha(p+k)}\cos(p+k)\theta. \end{cases} \tag{61}$$

Proof of (60)–(61). As for (54)–(55), we have

$$\begin{aligned} &4(\Lambda_2(\cos p\theta)\sin k\theta \mid \cos m\theta) \\ &\quad = 2(-[\cos m\theta, \sin p\theta] \mid \cos k\theta) - 2([\cos m\theta, \cos k\theta] \mid \sin p\theta) \\ &\qquad - 2([\sin k\theta, \cos p\theta] \mid \cos m\theta) \\ &\quad = (m-p)\,\alpha(k)\delta_k^{m+p} - (p+m)\,\alpha(k)\delta_p^{m+k} - (m+p)\,\alpha(k)\delta_m^{k+p} \\ &\qquad + (k+m)\,\alpha(p)\delta_k^{m+p} + (k-m)\,\alpha(p)\delta_p^{m+k} - (k+m)\,\alpha(p)\delta_m^{k+p} \\ &\qquad + (k+p)\,\alpha(m)\delta_k^{m+p} + (k+p)\,\alpha(m)\delta_p^{m+k} + (k-p)\,\alpha(m)\delta_m^{k+p}. \end{aligned}$$

□

5. Commuting with the Hilbert transform, torsionless and antisymmetry

We prove that Γ_1 of last section is characterized by torsionless condition, commutation with J and antisymmetry condition.

We put ourselves on $\mathcal{G} = \mathrm{diff}(S^1)$. *Then having discussed the properties of* Γ *on* $\mathrm{diff}(S^1)$, *we take the orthogonal projection* $\pi : \mathrm{diff}(S^1) \to V$ *and we define, for* $v \in \mathrm{diff}(S^1)$, *the operator*

$$\Gamma_{\mathrm{proj}}(v) = \pi \circ \Gamma(v)_{|V} \tag{62}$$

where $\Gamma(v)_{|V}$ *denotes the restriction of* $\Gamma(v)$ *to the subspace* V. In fact, the metric on the linear subspace V of $\mathrm{diff}(S^1)$ will determine the curvature of the quotient space.

Torsionless condition and commutation with J *on* $\mathrm{diff}(S^1)$. In the following lemma, we characterize Γ when

$$\begin{array}{ll}\text{(i) } \Gamma(u)v - \Gamma(v)u = [u, v] \quad \text{(torsionless condition)} & \\ \text{(ii) } \Gamma(u)Jv = J\Gamma(u)v & \end{array} \quad \text{for } v \in \mathrm{diff}(S^1). \tag{63}$$

Notice that when $u = 1$ and v is in the subspace of $\mathcal{G}$ generated by $\{\cos k\theta, \sin k\theta\}_{k\geq 1}$, then with $\Gamma(1)u = u'$, $\Gamma(v)(1) = 0$ and $\Gamma(1)Jv = (Jv)' = Jv'$, the conditions (63) are satisfied. We consider the case where u and v are in the subspace generated by $\{\cos k\theta, \sin k\theta\}_{k\geq 1}$. Since $[\cos p\theta, \sin k\theta]$ is expressed in terms of cos, we put for p, $k \geq 1$,

$$\begin{aligned}\Gamma(\cos p\theta)\sin k\theta = \beta_p(k)\cos(p+k)\theta + \gamma_p(k)1_{p\geq k+1}\cos(p-k)\theta \\ + \mu_p(k)1_{k\geq p+1}\cos(k-p)\theta + a_k\delta_k^p \cos k\theta.\end{aligned} \tag{64}$$

Lemma. *Assume that* Γ *satisfies* (63)–(64); *then*

$$\begin{cases}\Gamma(\cos p\theta)\sin k\theta = \beta_p(k)\cos(p+k)\theta + \gamma_p(k)1_{p\geq k+1}\cos(p-k)\theta \\ \qquad\qquad +\mu_p(k)1_{k\geq p+1}\cos(k-p)\theta + a_k\,\delta_k^p, \\ \Gamma(\cos p\theta)\cos k\theta = -\beta_p(k)\sin(p+k)\theta - \gamma_p(k)1_{p\geq k+1}\sin(p-k)\theta \\ \qquad\qquad -\mu_p(k)1_{k\geq p+1}\sin(k-p)\theta,\end{cases} \tag{65a}$$

$$\begin{cases}\Gamma(\sin p\theta)\cos k\theta = \beta_p(k)\cos(p+k)\theta \\ \qquad\qquad +(\mu_p(k) - (p+k))\,1_{k\geq p+1}\cos(k-p)\theta \\ \qquad\qquad +\gamma_p(k)\,1_{p\geq k+1}\cos(p-k)\theta + (a_k - k)\delta_k^p, \\ \Gamma(\sin p\theta)\sin k\theta = \beta_p(k)\sin(p+k)\theta \\ \qquad\qquad +(\mu_p(k) - (p+k))\,1_{k\geq p+1}\sin(k-p)\theta \\ \qquad\qquad +\gamma_p(k)\,1_{p\geq k+1}\sin(p-k)\theta,\end{cases} \tag{65b}$$

where

$$\begin{aligned}-\beta_p(k) + \beta_k(p) &= \frac{1}{2}(p-k), \\ -\gamma_p(k) + \mu_k(p) &= \frac{1}{2}(k+p) \quad \textit{for } p \geq k+1.\end{aligned} \tag{66}$$

In particular,

$$\textit{if } \gamma_p(k)1_{p\geq k+1} = 0, \quad \textit{then } \mu_p(k) = \frac{1}{2}(k+p) \ \textit{ for } k \geq p+1. \tag{67}$$

Proof. From (63), and since $\cos k\theta = -J\sin k\theta$, we obtain $\Gamma(\cos p\theta)\cos k\theta$. With (63) (i) (torsionless condition), $\Gamma(\cos p\theta)\cos k\theta = \Gamma(\cos k\theta)\cos p\theta + [\cos p\theta, \cos k\theta]$. This gives the condition (66). With the torsionless condition, and with (64), we calculate

$$\Gamma(\sin p\theta)\cos k\theta = \Gamma(\cos k\theta)\sin p\theta + [\sin p\theta, \cos k\theta].$$

With (63) (ii) (commutation with J), we find $\Gamma(\sin p\theta)\sin k\theta$. The torsionless condition on $\Gamma(\sin p\theta)\sin k\theta$ yields again the conditions (66). □

Lemma. *Assume that Γ satisfies the conditions of the previous lemma, i.e., it is given by (65)–(66); then, for $p \geq 1$, the operator $\Gamma(\cos p\theta)^2 + \Gamma(\sin p\theta)^2$ is diagonal in the trigonometric basis $\{\cos k\theta, \sin k\theta\}_{k\geq 1}$ if and only if $\gamma_p(k)\,1_{p\geq k+1} = 0$. In that case, we have, for $k \geq 1$,*

$$\begin{aligned}\left[\Gamma(\cos p\theta)^2 + \Gamma(\sin p\theta)^2\right]\sin k\theta \\ = -[(2p+k)\beta_p(k) + (p+k)\beta_p(k-p)\,1_{k\geq p+1}]\sin k\theta. \quad (68)\end{aligned}$$

In particular, if $\Gamma = \Gamma_1$ or $\Gamma = \Lambda_2$, then $\left[\Gamma(\cos p\theta)^2 + \Gamma(\sin p\theta)^2\right]$ is a diagonal operator.

Proof.

$$\begin{cases}\Gamma(\cos p\theta)\sin k\theta \\ \quad = \sum_{r\geq 1}\left[\beta_p(k)\delta_r^{p+k} + \gamma_p(k)\delta_r^{p-k} + \mu_p(k)\delta_r^{k-p}\right]\cos r\theta + a_k\delta_k^p, \\ \Gamma(\cos p\theta)\cos r\theta \\ \quad = -\sum_{j\geq 1}\left[\beta_p(r)\delta_j^{p+r} + \gamma_p(r)\delta_j^{p-r} + \mu_p(r)\delta_j^{r-p}\right]\sin j\theta.\end{cases}$$

Thus

$$\begin{aligned}\Gamma(\cos p\theta)^2\sin k\theta = -\sum_{j\geq 1}\sum_{r\geq 1}&\left[\beta_p(k)\delta_r^{p+k} + \gamma_p(k)\delta_r^{p-k} + \mu_p(k)\delta_r^{k-p}\right] \\ &\times\left[\beta_p(r)\delta_j^{p+r} + \gamma_p(r)\delta_j^{p-r} + \mu_p(r)\delta_j^{r-p}\right]\sin j\theta\end{aligned}$$

and

$$\begin{aligned}\Gamma(\sin p\theta)^2\sin k\theta = \sum_{j\geq 1}\sum_{r\geq 1}&\left[\beta_p(k)\delta_r^{p+k} + \gamma_p(k)\delta_r^{p-k} + (\mu_p(k) - (p+k))\delta_r^{k-p}\right] \\ &\times\left[\beta_p(r)\delta_j^{p+r} + \gamma_p(r)\delta_j^{p-r} + (\mu_p(r) - (p+r))\delta_j^{r-p}\right]\sin j\theta.\end{aligned}$$

Adding, we find

$$
\begin{aligned}
&[\Gamma(\cos p\theta)^2 + \Gamma(\sin p\theta)^2] \sin k\theta \\
&\qquad = [k(p+k) - k\mu_p(k) - (p+k)\mu_p(k-p)]1_{k\geq 2p+1} \sin(k-2p)\theta \\
&\qquad\quad - (p+k)\gamma_p(k-p)1_{2p\geq k+1} \sin(2p-k)\theta \\
&\qquad\quad - [(2p+k)\beta_p(k) + (p+k)\beta_p(k-p)] \sin k\theta.
\end{aligned}
$$

We proceed in the same way with $[\Gamma(\cos p\theta)^2 + \Gamma(\sin p\theta)^2] \sin k\theta$,

$$
\begin{aligned}
&\big(\big[\Gamma(\cos p\theta)^2 + \Gamma(\sin p\theta)^2\big] \cos k\theta \,\big|\, \cos j\theta\big) \\
&\qquad\qquad = \big(\big[\Gamma(\cos p\theta)^2 + \Gamma(\sin p\theta)^2\big] \sin k\theta \,\big|\, \sin j\theta\big).
\end{aligned}
$$

□

Corollary 1. *For $\Gamma = \Gamma_1$, $p \geq 0$, $k \geq 2$,*

$$
\begin{aligned}
&2\Gamma(\cos p\theta)^2 \sin k\theta + 2\Gamma(\sin p\theta)^2 \sin k\theta \\
&\qquad = -(p+k)^2 \frac{\alpha(k-p)}{\alpha(k)} 1_{k\geq p+2} \sin k\theta - \frac{(2p+k)^2\alpha(k)}{\alpha(p+k)} \sin k\theta. \qquad (69)
\end{aligned}
$$

Moreover $\sum_{p\geq 2} \Gamma\big(\cos p\theta/\sqrt{\alpha(p)}\big)^2 + \Gamma\big(\sin p\theta/\sqrt{\alpha(p)}\big)^2$ is a diagonal operator. The coefficients on the diagonal are given by

$$
\lambda_k = \frac{1}{2}\sum_{p\geq 2}\left[-(p+k)^2 \frac{\alpha(k-p)}{\alpha(p)\alpha(k)} 1_{k\geq p+2} - \frac{(2p+k)^2\alpha(k)}{\alpha(p)\alpha(p+k)}\right] \qquad (70)
$$

where the series converge.

$$
\begin{cases}
[\Gamma(\cos\theta)^2 + \Gamma(\sin\theta)^2] \sin k\theta = -(k^2-2)\sin k\theta \\
\Gamma(1)^2 \sin k\theta = -k^2 \sin k\theta
\end{cases}
\qquad \text{for } k \geq 2. \qquad (71)
$$

In the same order of idea, looking for diagonal operators, we have

Lemma. *With the assumptions* (65)–(66) *on Γ, the operator*

$$
\Phi_p \;:\; u \to [\Gamma(\cos p\theta)u, \cos p\theta] + [\Gamma(\sin p\theta)u, \sin p\theta] \qquad (72)
$$

is diagonal in the trigonometric basis $\{\cos k\theta, \sin k\theta\}_{k\geq 1}$ if and only if $\gamma_p(k)\, 1_{p\geq k+1} = 0$. In that case we have, for $k \geq 1$, $\Phi_p(\sin k\theta) = \lambda_p^\Phi(k) \sin k\theta$ and $\Phi_p(\cos k\theta) = \lambda_p^\Phi(k) \cos k\theta$, where

$$
\lambda_p^\Phi(k) = -\frac{1}{2}(p+k)(2p-k)\, 1_{k\geq p+1} + (2p+k)\beta_p(k) - k\, a_k \delta_k^p. \qquad (73)
$$

Proof.

$$
\begin{aligned}
&[\Gamma(\cos p\theta)\sin k\theta, \cos p\theta] \\
&\quad = \frac{1}{2}\sum_{r\geq 1, j\geq 1}[\beta_p(k)\delta_r^{p+k} + \gamma_p(k)\delta_r^{p-k} + \mu_p(k)\delta_r^{k-p}] \\
&\qquad\qquad \times [(r-p)\delta_j^{r+p} + (r+p)\delta_j^{r-p} - (r+p)\delta_j^{p-r}]\sin j\theta \\
&\quad + a_k\,\delta_k^p\,[1, \cos p\theta],
\end{aligned}
$$

$$
\begin{aligned}
&[\Gamma(\sin p\theta)\sin k\theta, \sin p\theta] \\
&\quad = \frac{1}{2}\sum_{r\geq 1, j\geq 1}[\beta_p(k)\delta_r^{p+k} + \gamma_p(k)\delta_r^{p-k} + (\mu_p(k) - (p+k))\delta_r^{k-p}] \\
&\qquad\qquad \times [(p-r)\delta_j^{r+p} + (r+p)\delta_j^{r-p} - (r+p)\delta_j^{p-r}]\sin j\theta.
\end{aligned}
$$

Adding, we see that the terms corresponding to $j = k-2p$ and $j = 2p-k$ vanish if and only if $\mu_p(k) = \frac{1}{2}(p+k)$. This gives the condition on Φ_p to be diagonal. In that case, we calculate $\lambda_p^{\Phi}(k)$. □

Theorem. *We keep the assumptions* (65)–(66) *on* Γ. *For* $p \geq 1$, *the operator* $\mathbf{Z}_p$ *defined by*

$$
\mathbf{Z}_p = \Phi_p + \Gamma(\cos p\theta)^2 + \Gamma(\sin p\theta)^2 \tag{74}
$$

is diagonal and $\mathbf{Z}_p \sin k\theta = \lambda_p^Z(k)\sin k\theta$, $\mathbf{Z}_p\cos k\theta = \lambda_p^Z(k)\cos k\theta$ *with*

$$
\lambda_p^Z(k) = -1_{k\geq p+1}(p+k)\left[\frac{1}{2}(2p-k) + \beta_p(k-p)\right] - k\,a_k\delta_k^p. \tag{75}
$$

Corollary 1. *With* (75), *for* $p \geq 0$, *we assume that* $\beta_p(k) = \frac{(2p+k)\alpha(k)}{2\alpha(p+k)}$ *(notice that this value of* $\beta_p(k)$ *is in* Γ_1*). Then*

$$
\lambda_p^Z(k) = \lambda_p^R(k)\alpha(p) \;\; \text{with} \;\; \lambda_p^R(k) = -1_{k\geq p+1}\times(p+k)\frac{(2k-p)}{2\alpha(k)} - k\,a_k\frac{1}{\alpha(k)}\delta_k^p. \tag{76}
$$

Moreover,

$$
2\sum_{p\geq 0}\lambda_p^Z(k) = -\frac{13}{6} - \frac{2k(a_k-1)}{\alpha(k)}. \tag{77}
$$

Proof. For $p \geq 1$, $\beta_p(k-p) = \frac{(p+k)\alpha(k-p)}{2\alpha(k)}$. To calculate the sum, we remark that

$$
\sum_{1\leq p\leq k-1}\frac{(p+k)(2k-p)}{\alpha(k)} = \frac{13}{6}k(k^2-1) - \frac{2k(k-1)}{\alpha(k)};
$$

then, we add the two terms corresponding to $p = k$ and $p = 0$. We see that the corresponding operator is bounded and that

$$\text{if } a_k = 1, \text{ then } 2\sum_{p\geq 0} \lambda_p^Z(k) = -\frac{13}{6}.$$

We have $[\Gamma(\cos\theta)u, \cos\theta] + [\Gamma(\sin\theta)u, \sin\theta] + \Gamma(\cos\theta)^2 u + \Gamma(\sin\theta)^2 u = 0$ and $[\Gamma(1)u, 1] + \Gamma(1)^2 u = 0$. □

Theorem (of unicity). *Let* Γ *be given by* (64),

$$\Gamma(\cos p\theta)\sin k\theta = \beta_p(k)\cos(p+k)\theta + \gamma_p(k)\,1_{p\geq k+1}\cos(p-k)\theta \\ + \mu_p(k)\,1_{k\geq p+1}\cos(k-p)\theta + a_k\delta_k^p.$$

Assume that $\Gamma(u)v - \Gamma(v)u = [u,v]$ *and* $\Gamma(u)Jv = J\Gamma(u)v$. *Then the condition*

$$(\Gamma(v)u \mid w) = \epsilon(\Gamma(v)w \mid u) \quad \textit{with } \epsilon^2 = 1 \tag{78}$$

implies that $\epsilon = -1$ *and*

$$\beta_p(k) = \frac{(2p+k)\alpha(k)}{2\alpha(p+k)}, \quad \textit{and for } k \geq p+1,\ \gamma_k(p) = 0,\ \mu_p(k) = \frac{p+k}{2}. \tag{79}$$

Proof. We define $H(u)w$ with $(H(u)w \mid v) = (\Gamma(v)u \mid w)$. We calculate $H(u)w$ in the trigonometric basis,

$$\begin{aligned} H(\cos p\theta)\sin k\theta &= -\frac{\beta_{k-p}(p)\alpha(k)}{\alpha(k-p)}\,1_{k\geq p+1}\cos(k-p)\theta \\ &\quad -\frac{\gamma_{k+p}(p)\alpha(k)}{\alpha(k+p)}\cos(k+p)\theta \\ &\quad -\frac{\mu_{p-k}(p)\alpha(k)}{\alpha(p-k)}\,1_{p\geq k+1}\cos(p-k)\theta, \end{aligned}$$

$$\begin{aligned} H(\sin k\theta)\cos p\theta &= \frac{\beta_{p-k}(k)\alpha(p)}{\alpha(p-k)}\,1_{p\geq k+1}\cos(p-k)\theta \\ &\quad +\frac{\gamma_{k+p}(k)\alpha(p)}{\alpha(k+p)}\cos(k+p)\theta \\ &\quad +\frac{\mu_{k-p}(k)\alpha(p)}{\alpha(k-p)}\,1_{k\geq p+1}\cos(k-p)\theta. \end{aligned}$$

The relation $H(\sin k\theta)\cos p\theta = \epsilon H(\cos p\theta)\sin k\theta$ gives

$$\begin{aligned} \gamma_{p+k}(k)\alpha(p) &= -\epsilon\gamma_{k+p}(p)\alpha(k) \\ \beta_{p-k}(k)\alpha(p) &= -\epsilon\mu_{p-k}(p)\alpha(k) \quad \text{for } p \geq k+1. \end{aligned} \tag{i}$$

In a similar way, we have

$$\begin{aligned} H(\cos p\theta)\cos k\theta &= \frac{\beta_{k-p}(p)\alpha(k)}{\alpha(k-p)}\,1_{k\geq p+1}\,\sin(k-p)\theta \\ &+ \frac{\gamma_{k+p}(p)\alpha(k)}{\alpha(k+p)}\sin(k+p)\theta \\ &+ \frac{\mu_{p-k}(p)-(2p-k)\alpha(k)}{\alpha(p-k)}\,1_{p\geq k+1}\,\sin(p-k)\theta. \end{aligned}$$

The condition $H(\cos p\theta)\cos k\theta = \epsilon H(\cos k\theta)\cos p\theta$ gives

$$\begin{aligned} &\gamma_{p+k}(p)\alpha(k) = \epsilon\gamma_{p+k}(k)\alpha(p) \\ &\beta_{k-p}(p)\alpha(k) = \epsilon(\mu_{k-p}(k)-(2k-p))\alpha(p) \quad \text{for } k\geq p+1. \end{aligned} \tag{ii}$$

It implies that $\gamma = 0$. The second equation in (ii) can be written as

$$\beta_{p-k}(k)\alpha(p) = \epsilon(\mu_{p-k}(p)-(2p-k))\,\alpha(k)$$

for $p \geq k+1$. The comparison with (i) gives

$$\mu_k(p) = \frac{k+p}{2} \quad \text{for } p\geq k+1.$$

Replacing in $\beta_k(p)\,\alpha(k+p) = \epsilon(\mu_k(k+p)-(2k+p))\,\alpha(p)$, we find

$$\beta_k(p) = -\epsilon\frac{(2k+p)\alpha(p)}{2\alpha(k+p)}.$$

Then the condition $\beta_k(p)-\beta_p(k) = \frac{1}{2}(p-k)$ determines $\epsilon = -1$. □

Part II. Stochastics on H \ Diff (S^1) and integration by parts formula

We identify the quotient space $\operatorname{diff}(S^1)/\operatorname{su}(1,1)$ with the subspace V of $\operatorname{diff}(S^1)$ generated by $\{\cos k\theta, \sin k\theta\}_{k\geq 2}$. The following is valid when Γ is the Levi-Civita connection. Let $(\epsilon_j)_{j\geq 2}$ be the orthonormal basis of V defined by (22). Let $(x_j(t))_{j\geq 0}$ be independent Brownian motions. We put

$$x(t) = \sum_j \epsilon_j x_j(t). \tag{80}$$

We denote '$*d$' the Stratonovitch differential and 'o' the composition of maps. On $\mathcal{H} \setminus \mathrm{Diff}(S^1)$, let $\gamma(t)$ be the solution of

$$d\gamma(t) = \sum_{j\geq 2} (\epsilon_j \, o \, \gamma(t)) * dx_j(t) \quad \text{with } \gamma(0) = \mathrm{Id}. \tag{81}$$

We denote

$$*dx(t) \, o \, \gamma(t) = \sum_j (\epsilon_j \, o \, \gamma(t)) * dx_j(t). \tag{82}$$

For fixed j, then $\epsilon_j x_j(t)$ is a process in $\mathrm{diff}(S^1)$. More generally, consider random vectors $(y_j(t))_j$ with $y_j(t) \in \mathrm{diff}(S^1)$ for any j. Let

$$y(t) = \sum_{j\geq 2} y_j(t). \tag{83}$$

We put

$$*dy(t) \, o \, \gamma(t) = \sum_j *dy_j(t) \, o \, \gamma(t). \tag{84}$$

For the Levi-Civita connection Γ, for any h in V, consider the parallel transport given by (see [1])

$$d\Omega(t)h = \Gamma(* \, dx(t)) \, \Omega(t) \, h = \sum_j \Gamma(\epsilon_j) \, \Omega(t)h \, * \, dx_j(t). \tag{85}$$

With the notations of (14)–(17), we consider the process

$$\gamma_\mu(t) = \exp(\mu\Omega(t)h) \, o \, \gamma(t). \tag{86}$$

Theorem (Integration by parts). *Let $\Omega(\bullet)h$ be a solution of* (85). *For $F : X \to R$, we define $D_h^\Gamma F(\gamma(\bullet))$ as in* (18), *then* (19) *holds.*

Proof. It is an adaptation of Fang's proof [11]. Since the adaptation is not straightforward, we give the details and we divide the proof into three steps.

Step 1. Consider the process $\gamma_\mu(t) = \exp(\mu\Omega(t)h) \, o \, \gamma(t)$ as in (86). We construct a *tangent process* $\widetilde{\gamma}_\mu(t)$ such that (i), (ii) and (iii) are fulfilled,

$$\begin{aligned} &\text{(i) } \widetilde{\gamma}_0(t) = \gamma(t), \\ &\text{(ii) } d\widetilde{\gamma}_\mu(t) = (dy^\mu(t) + \mu \, z(t) \, dt) \, o \, \widetilde{\gamma}_\mu(t) \quad \text{with } \widetilde{\gamma}_\mu(0) = \mathrm{Id}, \\ &\text{(iii) } \left.\frac{d}{d\mu}\right|_{\mu=0} \widetilde{\gamma}_\mu = \left.\frac{d}{d\mu}\right|_{\mu=0} \gamma_\mu, \end{aligned} \tag{87}$$

where $y^\mu(t) = \int_0^t \exp(\mu\Gamma(\Omega(s)h))\,dx(s)$. Since the operator $\Gamma(\omega(s)h)$ is antisymmetric from V to V, then $y^\mu(t)$ is a Brownian motion on $\mathrm{diff}(S^1)$.

Construction of the tangent process $\widetilde{\gamma}_\mu$ *and the expressions of* $z(s)$ *and* $dy^\mu(s)$ *in (ii).*

Proposition 1. *Let*

$$M(t) = \left(\frac{d}{d\mu}\gamma_\mu\right)\bigg|_{\mu=0}; \tag{88}$$

then $M(t)$ *satisfies*

$$dM(t) = *dy(t)\,o\,\gamma(t) + *dx(t)\,o\,M(t) \tag{89}$$

and the differential of $y(t)$ *in Itô's form is*

$$dy(t) = \Gamma(\Omega(t)h)\,dx(t) + \Psi\,\Omega(t)h\,dt. \tag{90}$$

Proof. Notice that $dx(t) \in \mathrm{diff}(S^1)$ and $\Gamma(\Omega(t)h)$ is an antisymmetric operator on $\mathrm{diff}(S^1)$. Taking the stochastic derivative of $\gamma_\mu(t)$ in (17), we obtain

$$\begin{aligned} d\gamma_\mu(t) &= *d(\exp(\mu\Omega(t)h))\,o\,\exp(-\mu\Omega(t)h)\,o\,\gamma_\mu(t) \\ &\quad + (\exp(\mu\Omega(t)h))'\,o\,*\,d\gamma(t) \\ &= \big(\mu * d(\Omega(t)h) + \text{terms in } \mu^j \text{ with } j\geq 2\big)\,o\,\gamma_\mu \\ &\quad + (\exp(\mu\Omega(t)h))'\,o\,*\,d\gamma(t). \end{aligned} \tag{91}$$

On the other hand, because of (81),

$$\begin{aligned} (\exp(\mu\Omega(t)h))'\,o\,*\,d\gamma(t) &= \sum_j (\exp(\mu\Omega(t)h))'\,o\,\epsilon_j\,o\,\gamma(t)\ *\,dx_j(t) \\ = \sum_j (I+\mu(\Omega(t)h)')\,o\,\epsilon_j\,o\,&\exp(-\mu\Omega(t)h)\,o\,\gamma_\mu(t)\ *\,dx_j(t) + \text{terms in } \mu^j, \\ & j\geq 2. \end{aligned} \tag{92}$$

We differentiate this last identity with respect to μ and take $\mu = 0$, it gives

$$\begin{aligned} &\frac{d}{d\mu}\bigg|_{\mu=0}\big[(\exp(\mu\Omega(t)h))'\,o\,*\,d\gamma(t)\big] \\ &\quad = \sum_j (\Omega(t)h))'\,o\,\epsilon_j - \epsilon_j'\,o\,\Omega(t)h)\,o\,\gamma(t)\ *\,dx_j(t) + \epsilon_j\,o\,M(t)\ *\,dx_j(t) \\ &\quad = \sum_j \big[\Omega(t)h,\epsilon_j\big]\,o\,\gamma(t)\ *\,dx_j(t) + \epsilon_j\,o\,M(t)\ *\,dx_j(t), \end{aligned} \tag{93}$$

The bracket $[\Omega(t)h, \epsilon_j]$ is given by $[\Omega(t)h, \epsilon_j] = (\Omega(t)h))' \, o \, \epsilon_j - \epsilon_j' \, o \, \Omega(t)h$. Notice that in our case the Lie bracket is different from the one in [11]. We deduce that

$$dM(t) = \sum_j \big(\Gamma(\epsilon_j)\Omega(t)h + [\Omega(t)h, \epsilon_j]\big) \, o \, \gamma(t) * dx_j(t) + \sum_j \epsilon_j \, oM(t) * dx_j(t). \tag{94}$$

We put

$$\begin{aligned} dy_j(t) &= \big(\Gamma(\epsilon_j)\Omega(t)h + [\Omega(t)h, \epsilon_j]\big) \, * \, dx_j(t) \\ &= \Gamma(\Omega(t)h)\epsilon_j \, * \, dx_j(t), \end{aligned} \tag{95}$$

where the last equality is a consequence of

$$\Gamma(\epsilon_j)\Omega(t)h + [\Omega(t)h, \epsilon_j] = \Gamma(\Omega(t)h)\epsilon_j. \tag{96}$$

Compare with (49). The equation (95) is a Stratonovitch equation. Let

$$y(t) = \sum_j y_j(t). \tag{97}$$

We have in Stratonovitch differential

$$dy(t) = \sum_j \Gamma(\Omega(t)h)\epsilon_j \, * \, dx_j(t) \tag{98}$$

and we have

$$dM(t) = *dy(t) \, o \, \gamma(t) + *dx(t) \, o \, M(t). \tag{99}$$

From (95), we calculate Itô's stochastic differential of $y(t)$. The Itô contractions are given by

$$\begin{aligned} I_j dt &= \big\langle d(\Gamma(\epsilon_j)\Omega(t)h + [\Omega(t)h, \epsilon_j]), dx_j(t) \big\rangle \\ &= [\Gamma(\epsilon_j)\Gamma(\epsilon_j)\Omega(t)h + \phi(\epsilon_j)\Gamma(\epsilon_j)\Omega(t)h) \, dt \end{aligned} \tag{100}$$

and

$$\sum_j I_j \, dt = \Psi\Omega(t)h \, dt, \tag{101}$$

where Ψ and ϕ are given by (11)–(12). We see that Itô's differential of $y(t)$ is given by (90). This proves Proposition 1. □

Definition. For $t \geq 0$, let Q_t^μ be the solution of

$$dQ_t^\mu = \Gamma(\Omega(t)h)Q_t^\mu \, d\mu \quad \text{with} \;\; Q_t^0 = \mathrm{Id}_{\mathrm{diff}(S^1)}. \tag{102}$$

We have

$$\left.\frac{d}{d\mu}\right|_{\mu=0} Q_t^\mu = \Gamma(\Omega(t)h). \tag{103}$$

Since $\Gamma(\Omega(t)h)$ is antisymmetric on $\mathrm{diff}(S^1)$, then Q_t^μ is orthogonal with the scalar product. We define

$$y^\mu(t) = \int_0^t Q_s^\mu \, dx(s); \tag{104}$$

then $y^0(t) = x(t)$ and $y^\mu(t)$ is a Brownian motion on $\mathrm{diff}(S^1)$ from the orthogonality of Q_s: the Itô contraction

$$\langle dy^\mu(t), dy^\mu(t)\rangle = \sum_j \|Q_t \epsilon_j\|^2 dt = dt. \tag{105}$$

Proposition 2. *Let $\widetilde{\gamma}_\mu$ be the solution of the Stratonovitch equation*

$$d\widetilde{\gamma}_\mu = \big(*dy^\mu(t) + \mu \Psi\Omega(t)h\, dt\big)\, o\, \widetilde{\gamma}_\mu \tag{106}$$

and let

$$\widetilde{M} = \left.\frac{d}{d\mu}\right|_{\mu=0} \widetilde{\gamma}_\mu\,; \tag{107}$$

we have $\widetilde{\gamma}_0(t) = \gamma(t)$ since $y^0(t) = x(t)$. Moreover, $\widetilde{M} = M$.

Proof.

$$d\widetilde{M} = \left(d\left(\left.\frac{d}{d\mu}\right|_{\mu=0} y_\mu\right) + \Psi\Omega(t)h\, dt\right) o\, \gamma(t) + (*dx(t))\, o\, \widetilde{M}. \tag{108}$$

We have to verify that

$$dy(t) = d\left(\left.\frac{d}{d\mu}\right|_{\mu=0} y_\mu\right) + \Psi\Omega(t)h\, dt \tag{109}$$

or equivalently

$$d\left(\left.\frac{d}{d\mu}\right|_{\mu=0} y_\mu\right) = \Gamma(\Omega(t)h)\, dx(t). \tag{110}$$

To verify (110), we differentiate $dy^\mu(t) = Q_t^\mu \, dx(t)$ with respect to μ. It gives

$$d\left(\left.\frac{d}{d\mu}\right|_{\mu=0} y_\mu\right) = \left(\left.\frac{d}{d\mu}\right|_{\mu=0} Q_t^\mu\right) dx(t).$$

Since $M(t)$ and $\widetilde{M}(t)$ satisfy the same stochastic differential equation, we conclude that $\widetilde{M} = M$. □

Step 2. The Girsanov formula is written for the tangent process $\widetilde{\gamma}_\mu$. Let

$$T_\mu : \gamma \to \widetilde{\gamma}_\mu$$

and the *density* K_μ

$$K_\mu(\gamma)(t) = \exp\left(-\mu \int_0^t (z_\gamma(s) \mid dy_\mu(t)) - \mu^2 \int_0^t |z_\gamma(s)|^2_{H^{\frac{3}{2}}} ds\right)$$

with $z_\gamma(s) = \Psi\Omega(s)h$ and where $\Omega(s)$ is the parallel transport along γ. From Girsanov theorem, for $F : X \to R$,

$$E[(F\, o T_\mu) \times K_\mu] = E[F].$$

Step 3. We differentiate with respect to μ the previous formula,

$$E[D_h^\Gamma F] = -E\left[F \frac{d}{d\mu}\bigg|_{\mu=0} K_\mu\right].$$

Since $\frac{d}{d\mu}\big|_{\mu=0} K_\mu = -\int_0^t (z_\gamma(s) \mid dx(s))$, we obtain the integration by parts formula. □

References

1. H. Airault, P. Malliavin, "Quasi-invariance of Brownian measures on the group of circle homeomorphisms and infinite-dimensional Riemannian geometry", *J. Funct. Anal.* (2006).
2. H. Airault, J. Ren, "Modulus of continuity of the canonic Brownian motion "on the group of diffeomorphisms of the circle" ", *J. Funct. Anal.* **196**, 395–426 (2002).
3. H. Airault, P. Malliavin, Anton. Thalmaier, "Canonical Brownian motion on the space of univalent functions and resolution of Beltrami equations by a continuity method along stochastic flows", *J. Math. Pure Appl.* **83**, 955–1018 (2004).
4. H. Airault, "Riemannian connections and curvatures on the universal Teichmuller space", *C. R. Acad. Sci. Paris*, Aout 2005.
5. J. M. Bismut, *Large deviations and Malliavin Calculus*, Birkhäuser, Basel (1984).
6. M. J. Bowick, S. G. Rajeev, "String theory as the Kähler geometry of loop space", *Phys. Rev. Letter* **58**, no. 6 (1987).
7. M. J. Bowick, S. G. Rajeev, "The holomorphic geometry of closed bosonic string theory and $\mathrm{Diff}(S^1)/S^1$", *Nuclear Physics B* **293**, 348–384 (1987).
8. M. J. Bowick, A. Lahiri, "The Ricci curvature of $\mathrm{Diff}(S^1)/SL(2,R)$", *J. Math. Phys.* **29**, no. 9, 1979–1980 (1988).

9. A. B. Cruzeiro, P. Malliavin, "Renormalized differential geometry on path space: structural equation, curvature", *J. Funct. Anal.* **139**, no. 1, 119–180 (1996).
10. B. Driver, "Integration by parts and quasi-invariance for heat kernel measures on loop groups", *J. Funct. Anal.* **149**, 470–547 (1997).
11. S. Fang, "Integration by parts for heat measures over loop groups", *J. Math. Pures Appl.* **78**, 877–894 (1999).
12. S. Fang, P. Malliavin, "Stochastic analysis on the path space of a Riemannian manifold", *J. Funct. Anal.* **118**, no. 1, 249–274 (1993).
13. S. Fang, "Canonical brownian motion on the diffeomorphism group of the circle", *J. Funct. Anal.* (2002).
14. M. Gordina, P. Lescot, "Riemannian geometry on $\mathrm{Diff}(S^1)/S^1$", *J. Funct. Anal.* (2006).
15. D. K. Hong, S. G. Rajeev, "Universal Teichmüller space and $\mathrm{Diff}\, S^1/S^1$, *Commun. Math. Phys.* **135**, 401–411 (1991).
16. A. A. Kirillov, D. V. Yurev, "Kähler geometry of the infinite-dimensional homogeneous space $M = \mathrm{Diff}_+(S^1)/\mathrm{Rot}(S^1)$", Translated from *Funkt. Anal. i Ego Priloz.* **20**, no 4, 79–80 (1986).
17. P. Malliavin, "The canonic diffusion above the diffeomorphism group of the circle", *C. R. Acad. Sci. Paris*, Serie 1, 325–329 (1999).

HÉLÈNE AIRAULT
INSSET, 48, rue Raspail, 02100 Saint-Quentin, Aisne
and
LAMFA, UMR 6140 CNRS 33, rue Saint-Leu, 80039 Amiens, France
hairault@insset.u-picardie.fr

INVARIANT MEASURES FOR ORNSTEIN-UHLENBECK OPERATORS

HÉLÈNE AIRAULT (*INSSET, Saint-Quentin*),
PAUL MALLIAVIN (*Paris*)

We produce a proof susceptible of generalization of the following result: *the classical Ornstein-Uhlenbeck operator has for invariant measure the law* ν *at time* 1 *of the Brownian motion starting from* 0 *at time* 0. Let $\mathcal{M}$ be a Riemannian manifold. On $\mathcal{M}$, let Z be a vector field and ν be the law of a diffusion with infinitesimal generator $\triangle$. The condition $[\triangle, Z] = \lambda Z + \gamma\triangle$ where λ and γ are two constants permits to obtain that the semi-group $P_t = \exp(t(\triangle + cZ))$ has for invariant measure the transition probability associated to the semi-group $\exp(t\triangle)$.

1. The Ornstein-Uhlenbeck operator on a Berezinian space

Given a Kählerian manifold $\mathcal{M}$, of dimension n, let ω be its symplectic form, assume the existence of a globally defined *Kähler potential* K, that is K is a globally defined C^2 function such that $\partial\bar{\partial}K = i\omega$. Define a Berezinian measure as a probability measure μ of the form

$$\mu_c := \gamma \ \exp(-cK) \ (\omega)^{\wedge n}, \tag{1}$$

where c is a positive constant and where γ is a normalizing constant. Berezinian measures appear in the theory of representations of finite dimensional Lie groups; in infinite dimension, it has been recently discovered. See [3, Theorem (4.2.4)] that unitary representations of Virasoro algebra can been described in a suitably defined Berezinian context: the right hand side of (1) becomes in infinite dimension meaningless and the measure μ is then defined as the reversible invariant probability measure associated to the elliptic operator

$$\triangle - \frac{c}{2}\nabla K * \nabla \tag{2}$$

where $\triangle$ is the infinitesimal generator of the Brownian motion on $\mathcal{M}$ (that is $\frac{1}{2}$ the Laplace-Beltrami operator of classical differential geometry). See [7]. The *pole* will be the point m_0 where K reaches its minimum.

We denote by $\pi_s(m_0, dm)$ the law of the Brownian motion at time s, conditioned to start from m_0 at time $t = 0$. The question object of this work is when $\mu_c = \pi_s$ for a suitable choice of s? As the law of Brownian motion perturbed by a drift in infinite dimension is often constructible this identity will furnish an alternative route for construction of Berezinian measures in infinite dimension. This procedure was started by P. Malliavin in [7]. In the present, we discuss the finite dimensional setting. In [1], they consider real valued processes. Taking the law of a Levy Brownian motion at time 1, they obtain an invariant measure of the process and study the associated Dirichlet forms. In [9], the Ornstein-Uhlenbeck process was constructed in a Riemannian context. Here, we develop some properties of the Ornstein-Uhlenbeck operator on a finite dimensional complex Kählerian manifold. In particular, for the Poincaré disk (see for example [4]), we obtain that in the case of the Ornstein-Uhlenbeck process, we have to consider the radial part of the process and the drift at the point m is given by the Poincaré distance from m to the origin, this in some sense ties with [9]. The radial part can be studied with the projection method of [2].

2. The classical one-dimensional Ornstein-Uhlenbeck process

On the real line $\mathbb{R}$, consider the measure $\mu_t(x, dy)$ defined by

$$\int f(y)\,\mu_t(x,dy) = \frac{1}{\sqrt{2\pi}}\int_{-\infty}^{+\infty} f(e^{-t}x + \sqrt{1-e^{-2t}}y)e^{-y^2/2}\,dy \tag{3}$$

and denote

$$\widetilde{P}_t f(x) = \int f(y)\,\mu_t(x,dy). \tag{4}$$

We have

$$\int_{-\infty}^{+\infty} \widetilde{P}_t\, f(x)\,d\nu(x) = \int_{-\infty}^{+\infty} f(x)\,d\nu(x) \quad \text{with} \quad d\nu(x) = e^{-x^2/2}\,dx. \tag{5}$$

Thus $d\nu(x)$ is an invariant measure for the semi-group $\widetilde{P}_t$. By a change of variables,

$$\int f(y)\,\mu_t(x,dy) = \int f(y)\,\pi_t(x,y)\,e^{-y^2/2}\,dy$$

with

$$\pi_t(x,y) = \frac{1}{\sqrt{1-e^{-2t}}}\exp\left[-\frac{1}{2}\frac{(y-e^{-t}x)^2}{(1-e^{-2t})} + \frac{y^2}{2}\right]$$

and the density $\pi_t(x, y)$ satisfies

$$\widetilde{\Delta}\pi_t(x,y) = \frac{\partial}{\partial t}\pi_t(x,y) \quad \text{with} \quad \widetilde{\Delta} = \frac{1}{2}\left(-x\frac{\partial}{\partial x} + \frac{\partial^2}{\partial x^2}\right).$$

On the other hand, let P_t be the semi-group associated to the one dimensional Brownian motion on the real line,

$$P_t f(x) = (1/\sqrt{2\pi}) \int_{-\infty}^{+\infty} f(y) \exp(-(y-x)^2/2t)\, dy;$$

then the invariance condition (5) is the same as

$$(P_1(\widetilde{\triangle} f))(0) = 0. \tag{6}$$

The fact that (5) implies (6) is immediate since (5) implies

$$\int \left[\frac{\widetilde{P}_t f(y) - f(y)}{t} \right] \nu(dy) = 0 \quad \forall t \geq 0 \tag{7}$$

and

$$\int \widetilde{\triangle} f(y)\, \nu(dy) = \lim_{t \to 0} \int \left[\frac{\widetilde{P}_t f(y) - f(y)}{t} \right] \nu(dy) = 0. \tag{8}$$

To prove that (6) implies (5) is more delicate. We assume that (8) is true for $f = \widetilde{P}_s g$. Then we use the semi-group property for $\widetilde{P}_t$ to deduce that the condition $\frac{d}{dt}\Big|_{t=0} \int \widetilde{P}_t f(y)\, \nu(dy) = 0$ for $f = \widetilde{P}_s g$ implies that it holds $\frac{d}{dt}\Big|_{t=s} \int \widetilde{P}_t g(y)\, \nu(dy) = 0$. This is true for any s, thus (5) holds.

The previous proof likely extends to more general situations. In the following, our aim is to provide the algebraic framework for such a generalization. Therefore, to stay more simple, we leave aside the difficulties that arise from the discussions of the functions spaces on which the operators are well defined. We call test function any function for which the formula has a meaning. We leave to the reader to take care of the details. Moreover the method of path spaces as presented in [8], may provide a better framework for possible extensions than a rigorous development in functions spaces.

3. The Ornstein-Uhlenbeck process on an Euclidean space E

Denote $\triangle$ the infinitesimal generator of the Brownian motion on E:

$$\triangle = \frac{1}{2} \sum_k \partial_k^2$$

and denote by $\tilde{\triangle}$ the infinitesimal generator of the Ornstein-Uhlenbeck process. Then

$$\tilde{\triangle} = \triangle - \frac{1}{2} \sum_k x_k \partial_k.$$

As $\partial_k^2 x_k = 2\partial_k + x_k\partial_k^2$, we have the commutation property

$$[\tilde{\triangle}, \triangle] = \triangle\,, \tag{9}$$

commutation which implies the following commutations, for any integer $m \geq 1$,

$$[\tilde{\triangle}, \triangle^m] = m\triangle^m$$

and finally

$$[\tilde{\triangle}, \exp(\triangle)] = \frac{d}{dt}\Big|_{t=1} \exp(t\triangle) \tag{10}$$

Introducing the semi-group $P_t = \exp(t\triangle)$, we want to prove that for every test function

$$A := \Big(P_1(\tilde{\triangle} f)\Big)(0) = 0. \tag{11}$$

With (10), we have

$$A = \Big(\tilde{\triangle} P_1(f)\Big)(0) - \Big(\frac{d}{dt}\Big|_{t=1} \exp(t\triangle) f\Big)(0);$$

then using the fact that the localization at the point 0 of the two operators $\tilde{\triangle}$, $\triangle$ coincide, we get

$$A = \Big(\triangle P_1(f)\Big)(0) - \Big(\frac{d}{dt}\Big|_{t=1} \exp(t\triangle) f\Big)(0),$$

relation which implies (11). From (11), it results that ν, the law at time 1 of the Euclidean Brownian motion starting from 0 is an invariant measure for the process generated by $\tilde{\triangle}$.

4. Invariant measures under a commutator hypothesis

The identities (9), (10), (11) can be generalized as follows

Theorem 4.1. *Let $\triangle$ be the infinitesimal generator of a diffusion on a manifold $\mathcal{M}$. Consider $\widetilde{\triangle} = \triangle - cZ$ where Z is a vector field and c is a constant. We assume that*

$$[\triangle\,, Z] = \lambda\,\triangle + \gamma Z \tag{12}$$

where λ and γ are constants. We define $L = \lambda\,\triangle + \gamma Z$. Assume $c > 0$, $c + \gamma > 0$ and denote $\rho = (c + \gamma)$, consider the semi-group associated to L, then for every test function f

$$\left(\exp\left(\frac{1}{\rho}L\right)\tilde{\triangle} f\right)(m_0) = 0. \tag{13}$$

Proof.

$$[\tilde{\Delta}, \Delta] = (c\lambda + \gamma)\Delta - \gamma\tilde{\Delta} = cL, \quad L := \lambda\Delta + \gamma Z, \quad [\tilde{\Delta}, L] = (c+\gamma)L. \tag{14}$$

$$[\tilde{\Delta}, L] = \rho L,$$
$$[\tilde{\Delta}, L^2] = \tilde{\Delta}L^2 - L^2\tilde{\Delta} = \rho L^2 + L\tilde{\Delta}L - L^2\tilde{\Delta} = 2\rho L^2.$$

$$\left[\tilde{\Delta}, \exp\left(\frac{1}{\rho}L\right)\right] = \frac{d}{dt}\Big|_{t=\frac{1}{\rho}} \exp(tL). \tag{15}$$

$$[\widetilde{\Delta}, L^n] = n\rho L^n. \tag{16}$$

Introducing the semi-group $P_t = \exp(tL)$ the commutation (16) implies that

$$A := \left(\exp\left(\frac{1}{\rho}L\right)\tilde{\Delta}f\right)(m_0) = \left(\tilde{\Delta}P_{\frac{1}{\rho}}(f)\right)(m_0) - \left(\frac{d}{dt}\Big|_{t=\frac{1}{\rho}} \exp(tL)f\right)(m_0).$$

Using the fact that the localization at the point m_0 of the two operators $\tilde{\Delta}$, L coincide, we get

$$A = \left(LP_{\frac{1}{\rho}}(f)\right)(m_0) - \left(\frac{d}{dt}\Big|_{t=\frac{1}{\rho}} \exp(tL)f\right)(m_0),$$

relation which implies (13). □

Main Theorem. *With the assumptions of Theorem 4.1, let $P_t = \exp(tL)$ and $\pi_t(m_0, dx)$ be the transition probability associated to this semi-group, i.e.,*

$$(P_t f)(m_0) = \int \pi_t(m_0, dx)\, f(x);$$

then $\pi_{\frac{1}{\rho}}(m_0, dx)$ is an invariant measure for the process generated by $\widetilde{\Delta}$,

$$\int (\widetilde{P}_t f)(y)\, \pi_{\frac{1}{\rho}}(m_0, dy) = \int f(y)\, \pi_{\frac{1}{\rho}}(m_0, dy) \quad \forall t \geq 0.$$

For the proof, see the equivalence of (6) and (5).

5.

The following one dimensional elementary example has been considered in [2]. It is straightforward that

$$\left[\frac{\partial^2}{\partial x^2}, x\frac{\partial}{\partial x}\right] = 2\frac{\partial^2}{\partial x^2}$$

and the previous theorems apply. Consider the image by a map g of a one dimensional Brownian motion. Assume that g is differentiable invertible from $\mathbb{R}$ to an interval (a,b) in $\mathbb{R}$. For example, $g(\tau) = \tanh(\tau)$. We take on the interval (a,b) the metric $ds^2 = dx^2/\alpha(x)$ with $\alpha(x) = [g'(g^{-1}(x))]^2$ (see [2] p. 379). The Laplace-Beltrami operator associated to ds^2 is

$$\triangle_g = [g'(g^{-1}(x))]^2\frac{d^2}{dx^2} + g''(g^{-1}(x))\frac{d}{dx}$$

and $\triangle_g(f) = \triangle(fog)of^{-1}$. The associated semi-group P_t^g is given by

$$P_t^g(f)(x) = P_t(fog)og^{-1}(x);$$

the probability of transition is

$$\pi_t^g(x,dz) = \frac{1}{2\pi t}\,\frac{1}{g'(g^{-1}(z))}\exp\left(-\frac{(g^{-1}(z)-g^{-1}(x))^2}{2t}\right)dz.$$

We put $Z = \alpha(x)d/dx$, assume that $[\triangle_g, Z] = \lambda\triangle_g + \gamma Z$ where λ and γ are constants. Then we must have $\gamma = 0$ and

$$\alpha(x) = \frac{\lambda}{2}g^{-1}(x)g'(g^{-1}(x)) + cg'(g^{-1}(x))$$

where c is a constant. Moreover, let $\widetilde{\triangle}_g = \triangle_g - cZ$, then the process starting from x and generated by $\widetilde{\triangle}_g$ has $\pi_t^g(x,dz)$ for invariant measure with $t = 1/\rho$. For $\lambda = 2$ and $x = g(\tau) = \tanh(\tau)$, we have

$$\tau = g^{-1}(x) = \frac{1}{2}\log\left[\frac{1+x}{1-x}\right] \quad \text{and} \quad \alpha(x)\frac{d}{dx} = \tau\frac{d}{d\tau} + c\frac{d}{d\tau}.$$

With $|z| = x$, the coefficient τ of $d/d\tau$ is the hyperbolic distance from 0 to z.

References

1. L. Accardi, V. Bogachev, "The Ornstein-Uhlenbeck process associated with the Levy Laplacian and its Dirichlet form", *Probab. Mat. Statist.* **17** no. 1, Acta Univ. Wratislav, n 1928, 95–114 (1997).
2. H. Airault, "Projection of the infinitesimal generator of a diffusion", *J. Funct. Anal.* **85** no. 2 (1989).
3. H. Airault, P. Malliavin, "Unitarizing probability measures for representation of Virasoro algebra", *J. Math. Pures Appl.* (9) 80, no. 6, 627–667 (2001).
4. H. Airault, "Stochastic analysis on finite dimensional Siegel disks, approach to the infinite dimensional Siegel disk and upper half-plane", *Bull. Sc. Math.* **128**, 605–659 (2004).
5. A. B. Cruzeiro, P. Malliavin, "Non perturbative construction of invariant measure through confinement by curvature", *J. Math. Pures Appl.* (9) 77, no. 6, 527–537 (1998).
6. B. Gaveau, J. Vauthier, "Annulations et calculs infinitesimaux de laplaciens pour un fibré non integrable", *Bull. Sci. Math* **100** no. 4, 353–368 (1976).
7. P. Malliavin, "The canonic diffusion above the diffeomorphism group of the circle", *C. R. Acad. Sc. Paris Ser. 1 Math. 329* no. 4, 325–329 (1999).
8. P. Malliavin, "Itô atlas", to appear in Proceedings of Abel conference, Oslo (2005), Springer.
9. D. Stroock, "The Ornstein-Uhlenbeck process on a Riemannian manifold", in *First International Congress of Chinese Mathematicians (Beijing, 1998)*, Amer. Math. Soc., Providence, RI, 2001, pp. 11–23.

HÉLÈNE AIRAULT
INSSET, Université de Picardie, 48, rue Raspail, 02100 Saint-Quentin, France
hairault@insset.u-picardie.fr

PAUL MALLIAVIN
10, rue Saint-Louis en l'ile, 75004, Paris, France
sli@ccr.jussieu.fr

BACKWARD STOCHASTIC DIFFERENTIAL EQUATIONS WITH RESPECT TO MARTINGALES*

AbdulRahman Al-Hussein (*Al-Qassim University*)

This paper is devoted to proving the existence and uniqueness of the solutions of backward stochastic differential equations driven by infinite dimensional martingales.

1. Introduction

Backward stochastic differential equations (BSDEs in short) have been widely studied over the last decade. These equations take usually the form (1) below. The appearance of such equations was first in the work of Bismut in [5] and later in the joint work of Pardoux and Peng in [12]. The main reason of studying such type of equations is to involve them in some mathematical problems; for instance in theory of PDEs, stochastic control and in finance.

In [2] we studied those BSDEs driven by a Wiener process on a Hilbert space H. The solutions of such equations were required to be adapted to the filtration generated by this driving Wiener process, i.e., the Wiener filtration. The question arising now is whether we are able to deal with such sort of BSDEs with a given arbitrary filtration, not necessary the Wiener filtration; for example the filtration $\mathcal{F}_t(M) = \sigma\{M(s),\ 0 \leq s \leq t\},\ t \geq 0$, where M is a square integrable cadlag martingale in H. In this work we will be concerned with giving answers to this question. Another example also could be the filtration generated by two independent cylindrical Wiener processes W_1 and W_2 on H. Note that if the terminal value ξ of the concerned BSDE is measurable with respect to $\mathcal{F}_T(W_1)$ and is independent of $\mathcal{F}_T(W_2)$, the solution (Y, Z) of the following equation can only be adapted to the filtration $\{\mathcal{F}_t(W_1) \bigvee \mathcal{F}_t(W_2),\ 0 \leq t \leq T\}$.

One advantage of working with a more general filtration than just the Wiener filtration is to enable us to study more equations than those focussed on just the Wiener filtration, e.g. as in [2] and [3]; cf. [4]. We shall study here on the space H the following *backward stochastic differential equation*

*Research supported by Al-Qassim University, project no. SR-D-006-003.

(BSDE):

$$\begin{cases} -dY(t) = f(t, Y(t), Z(t))\,dt - Z(t)\,dM(t) - dN(t), & 0 \le t \le T, \\ \quad Y(T) = \xi, \end{cases}$$

where M is a given square integrable martingale in H. For this we shall look for a triple (Y, Z, N) of adapted processes, square integrable and satisfy, for each $t \in [0, T]$, the integral form of this equation. Here N is a martingale required to be very strongly orthogonal (V.S.O.) to M, a notion will be given in Section 2. These equations are in fact backward stochastic differential equations driven by martingales.

The main purpose of this paper is to prove the existence and uniqueness of the solutions of this type of equations; see Theorem 3.1 in Section 3. These results generalize the work of El Karoui et al. [7] in finite dimensions, and also generalize the study of the usual BSDEs (see the equation (1) below), which was considered by Pardoux and Peng in [12]. For this issue, see the discussion following the proof of Theorem 3.1. Moreover, these BSDEs can be considered somehow as a generalization of some reflected BSDEs; see e.g. [8]. In fact, since the solutions of the reflected equations take usually values in $\mathbb{R}$, to make use of our result here, one should first reformulate the results in Theorem 3.1 for the case when the martingale M lies in the space H and Y lies in $\mathbb{R}$. This is however straightforward. Such type of BSDEs can also be applied in finance to construct the Föllmer-Schweiser strategy.

These BSDEs were applied also in [6] to study the approximation of the usual BSDEs as follows. Consider as in Pardoux and Peng [12] setting the following BSDE:

$$\begin{cases} -dY(t) = f(t, Y(t), Z(t))\,dt - Z(t)\,dW(t), & 0 \le t \le T, \\ \quad Y(T) = \xi, \end{cases} \tag{1}$$

with W being a genuine Wiener process (e.g. a Brownian motion in $\mathbb{R}$). By taking a martingale approximation of this W (see [6] for the definition), one obtains a sequence of equations, all of which are of the type (3). It was shown in [6] that this sequence of solutions actually converges to the solution started with. This result was done in fact for the finite dimensional case.

This paper is organized as follows. In Section 2 we recall some information on Hilbert space valued martingales and stochastic integration with respect to them. Section 3 contains the main results.

2. Basic elements of an infinite dimensional martingale and stochastic integration

A. Let $(\Omega, \mathcal{F}, \{\mathcal{F}_t\}_{t\geq 0}, \mathbb{P})$ be a complete filtered probability space, such that $\{\mathcal{F}_t\}_{t\geq 0}$ is right continuous. Fix $0 < T < \infty$. Let $\mathcal{R}$ be the algebra generated by elements of $\Omega \times (0, T]$ of the form $F \times (t, s]$, where $F \in \mathcal{F}_t$ and $t, s \in [0, T]$. Define $\mathcal{P}$ to be the σ-algebra generated by $\mathcal{R}$. The family of sets $\mathcal{P}$ is called the *predictable* σ-algebra. Let H be a separable Hilbert space. An H-valued process is said to be predictable if it is $\mathcal{P}/\mathcal{B}(H)$ measurable.

We shall write $H \otimes H$ for the tensor product of H with itself, denoting by $x \otimes y$ the tensor product of $x \in H$ and $y \in H$. Let $\mathcal{M}^2_{[0,T]}(H)$ denote the vector space of right continuous square integrable martingales $\{M(t), 0 \leq t \leq T\}$, taking values in H, that is $\sup_{t\in[0,T]} \mathbb{E}\left[|M(t)|^2_H\right] < \infty$. It is a separable Hilbert space with respect to the inner product $(M, N) \mapsto \mathbb{E}\left[\langle M(T), N(T)\rangle_H\right]$, if we agree to identify $\mathbb{P}$-equivalence classes. We say that the two elements M and N of $\mathcal{M}^2_{[0,T]}(H)$ are *very strong orthogonal (V.S.O.)* if $\mathbb{E}\left(M(u) \otimes N(u)\right) = \mathbb{E}\left(M(0) \otimes N(0)\right)$, for all $[0, T]$-valued stopping times u. For example, if moreover $N(0) = 0$, then M and N are V.S.O. if and only if $\mathbb{E}\left(M(u) \otimes N(u)\right) = 0$, for all such stopping times. Note that $\langle M(t), N(t)\rangle_H = \mathrm{tr}(M(t) \otimes N(t))$. This notion of orthogonality is stronger than the usual definition of strong and weak orthogonality. For further details see [10] and [9]. In [1] we considered notions of orthogonality (strong and very strong) slightly stronger than these used in this paper. In fact we found that our definitions here are more suitable; for instance, they are invariant under the shift by a constant.

Let us now recall the definition of *Doléans measure* associated with $|M|^2_H$. Define $d_{|M|^2_H}$ on elements $A = F \times (t, s]$ of $\mathcal{R}$ by $d_{|M|^2_H}(A) := \mathbb{E}\left[1_F(|M(s)|^2_H - |M(t)|^2_H)\right]$. This function can be extended uniquely to a measure α_M on $\mathcal{P}$. This measure is called the Doléans measure associated with $|M|^2_H$ (see [10] or [9]). Analogously, we associate on $\mathcal{P}$ the $H\hat{\otimes}_1 H$-valued σ-additive Doléans measure μ_M of $M \otimes M$. Here the space $H\hat{\otimes}_1 H$ is the completed nuclear tensor product, that is the completion of $H \otimes H$ for the nuclear norm. Recall that the linear form *trace*, denoted here by tr, is defined as the unique continuous extension to $H\hat{\otimes}_1 H$ of the mapping $x \otimes y \mapsto \langle x, y\rangle_H$.

For a square integrable martingale M we write $\langle M, M\rangle$ (or shortly $\langle M\rangle$) for the increasing Meyer process associated with the Doléans measure of the submartingale $|M|^2_H$, that is the unique (up to $\mathbb{P}$-equivalence) predictable, right continuous, increasing, real valued process, vanishing at zero such that

$|M|_H^2 - \langle M\rangle$ is a martingale. It exists since $|M|_H^2$ is a submartingale. We recall the following proposition from [10, Theorem 14.3.1, p. 167].

Proposition 2.1. *(1) There is one predictable $H\hat{\otimes}_1 H$-valued process $\mathcal{Q}_M$, defined up to α_M-equivalence such that for every $G \in \mathcal{P}$*

$$\mu_M(G) = \int_G \mathcal{Q}_M \, d\alpha_M.$$

Moreover, $\mathcal{Q}_M$ takes its value in the set of positive symmetric elements of $H\hat{\otimes}_1 H$ and

$$\operatorname{tr} \mathcal{Q}_M(\omega, t) = 1, \quad \alpha_M \text{ a.e.}$$

(2) The $H\hat{\otimes}_1 H$-valued process

$$\langle\langle M\rangle\rangle_t := \int_{(0,t]} \mathcal{Q}_M \, d\langle M\rangle$$

has finite variation, is predictable, admits μ_M as its Doléans measure, and is such that $M \otimes M - \langle\langle M\rangle\rangle$ is a martingale.

From this we conclude that M and N are V.S.O. if and only if $\langle\langle M, N\rangle\rangle = 0$, where $\langle\langle M, N\rangle\rangle$ is the unique (up to $\mathbb{P}$-equivalence) predictable $H\hat{\otimes}_1 H$-valued process with paths of finite variation vanishing at zero such that $M \otimes N - \langle\langle M, N\rangle\rangle$ is an $H\hat{\otimes}_1 H$-valued martingale.

To illustrate the above notions, let us for example consider the case of a 2-dimensional Brownian motion $B = (B^1, B^2)$, where B^1 and B^2 are two independent Brownian motions in $\mathbb{R}$. It is obvious that $\langle\langle B\rangle\rangle_t = \left(\begin{smallmatrix} t & 0 \\ 0 & t \end{smallmatrix}\right) =: t\, I_2$, and so $\langle B\rangle_t = 2t$ and $\mathcal{Q}_B = \frac{1}{2} I_2$. Moreover, μ_B is the product measure $(l \otimes \mathbb{P})\, I_2$ and $\alpha_B = (2\, l \otimes \mathbb{P})$, where l is the Lebesgue measure on $([0,T], \mathcal{B}([0,T]))$.

Denote by $L_1(H)$ the space of nuclear operators on H. It is known that elements of $H\hat{\otimes}_1 H$ can be identified with elements of $L_1(H)$. So we can let $\tilde{\mathcal{Q}}_M$ be the identification of $\mathcal{Q}_M$ in $L_1(H)$. Denote also by $L_2(H)$ the Hilbert space of all Hilbert-Schmidt operators from H to itself. We shall write that $G \in L_2^{\tilde{\mathcal{Q}}_M}(H)$ if $G \tilde{\mathcal{Q}}_M^{1/2} \in L_2(H)$.

B. Now we are ready to set the definition of stochastic integration with respect to elements of $\mathcal{M}^2_{[0,T]}(H)$. First, let $L^*(H; \mathcal{P}, M)$ be the space of processes Φ, the values of which are (possibly non-continuous) linear operators from H into itself with the following properties:

(i) the domain of $\Phi(\omega, t)$ contains $\tilde{\mathcal{Q}}_M^{1/2}(\omega, t)(H)$ for every (ω, t),

(ii) for every $h \in H$, the H-valued process $\Phi \circ \tilde{\mathcal{Q}}_M^{1/2}(h)$ is predictable,

(iii) for every $(\omega, t) \in \Omega \times (0, T]$, $\Phi(\omega, t) \circ \tilde{\mathcal{Q}}_M^{1/2}(\omega, t)$ is a Hilbert-Schmidt operator and

$$\int_{\Omega \times (0,T]} |\Phi \circ \tilde{\mathcal{Q}}_M^{1/2}|^2_{L_2(H)} \, d\alpha_M < \infty.$$

This space is complete with respect to the scalar product $(X, Y) \mapsto \int_{\Omega \times (0,T]} \mathrm{tr}(X \circ \tilde{\mathcal{Q}}_M \circ Y^*) \, d\alpha_M$; cf. [9, Proposition 22.2, p. 142]. See also [14].

Denote by $\mathcal{E}(L(H))$ the space of $\mathcal{R}$-simple processes and $\Lambda^2(H; \mathcal{P}, M)$ the closure of $\mathcal{E}(L(H))$ in $L^*(H; \mathcal{P}, M)$. It is therefore a Hilbert subspace of $L^*(H; \mathcal{P}, M)$. For a simple Φ of the form

$$\Phi = \sum_{i=1}^{n} 1_{F_i \times (r_i, s_i]} \, u_i, \quad u_i \in L(H; K), \; F_i \in \mathcal{F}_{r_i},$$

we define

$$\int_{(0,t]} \Phi \, dM = \sum_{i=1}^{n} 1_{F_i \times (r_i, s_i]} \, (u_i(M(s_i \wedge t)) - u_i(M(r_i \wedge t))), \quad t \in [0, T].$$

This gives an isometric linear mapping from $\mathcal{E}(L(H))$ into $\mathcal{M}^2_{[0,T]}(H)$, given by $\Phi \mapsto \int \Phi \, dM$. Extend this mapping to $\Lambda^2(H; \mathcal{P}, M)$. The image $\int \Phi \, dM$ of Φ in $\mathcal{M}^2_{[0,T]}(H)$ by this mapping is called the *stochastic integral of* Φ *with respect to* M. For such $\Phi \in \Lambda^2(H; \mathcal{P}, M)$ the stochastic integral $N = \int \Phi \, dM$ can easily be seen to satisfy the following two properties:

(1) $\langle N \rangle_t = \int_{(0,t]} \mathrm{tr}(\Phi \circ \tilde{\mathcal{Q}}_M \circ \Phi^*) \, d\langle M \rangle$,

(2) $\langle\langle N \rangle\rangle_t = \int_{(0,t]} (\Phi \circ \tilde{\mathcal{Q}}_M \circ \Phi^*) \, d\langle\langle M \rangle\rangle$,

for every $t \geq 0$.

The following representation property is due to [11]; see also [9, E. 8, p. 160].

Theorem 2.1. *Let* $M \in \mathcal{M}^2_{[0,T]}(H)$ *and*

$$\mathcal{H}_1 := \{\textstyle\int X \, dM : X \in \Lambda^2(H; \mathcal{P}, M)\} \subset \mathcal{M}^2_{[0,T]}(H).$$

Let $\mathcal{H}_2$ *be the orthogonal complement of* $\mathcal{H}_1$ *in* $\mathcal{M}^2_{[0,T]}(H)$. *Then every element of* $\mathcal{H}_2$ *is V.S.O. to every element of in* $\mathcal{H}_1$. *In particular, every*

$L \in \mathcal{M}^2_{[0,T]}(H)$ can be written uniquely as

$$L = \int X \, dM + N, \quad X \in \Lambda^2(H; \mathcal{P}, M), \; N \in \mathcal{H}_2 . \tag{2}$$

Note that since $M \in \mathcal{H}_1$, the martingales M and N are V.S.O.

3. Main results

A. This section contains the proof of the existence and uniqueness of the solution of the following type of BSDEs.

$$\begin{cases} -dY(t) = f(t, Y(t), Z(t)) \, dt - Z(t) \, dM(t) - dN(t), & 0 \le t \le T, \\ Y(T) = \xi. \end{cases} \tag{3}$$

The mappings f, ξ and M are required to satisfy the following conditions. First, let $\gamma > 0$ be fixed.

(H1) f is $\mathcal{P} \otimes \mathcal{B}(H) \otimes \mathcal{B}(L_2^{\tilde{Q}_M}(H))/\mathcal{B}(H)$-measurable.

(H2) $\exists \, k > 0$ such that $\forall \, y, y' \in H$, $\forall \, z, z' \in L_2^{\tilde{Q}_M}(H)$

$$|f(t,y,z) - f(t,y',z')|_H^2 \le k \left(|y - y'|_H^2 + |z - z'|^2_{L_2^{\tilde{Q}_M}(H)} \right),$$

uniformly in (t, ω).

(H3) $M \in \mathcal{M}^2_{[0,T]}(H)$, cadlag and $\langle\langle M \rangle\rangle_t = \int_0^t b(s) \, b(s)^* \, dc_s$, for some adapted continuous and increasing $\mathbb{R}_+$-valued process $\{c_s, \; s \ge 0\}$ such that $c_0 = 0$, and an $L_2(H)$-valued predictable process b. Here $b(s)^*$ is the adjoint of $b(s)$. In other words $\langle\langle M \rangle\rangle$ is absolutely continuous with respect to c.

(H4) $\mathbb{E}\left[\int_0^T e^{\gamma \, c_s} \, |f(s,0,0)|_H^2 \, dc_s \right] < \infty$.

(H5) $\mathbb{E}\left[e^{\gamma \, c_T} \, |\xi|_H^2 \right] < \infty$.

From (H3) it follows that, for all t,

$$\langle M \rangle_t = \int_0^t |b(s)|^2_{L_2(H)} \, dc_s$$

and

$$\tilde{\mathcal{Q}}_M(t) = \frac{b(t) \, b(t)^*}{|b(t)|^2_{L_2(H)}} .$$

Thus the condition (H2) becomes

$$|f(t,y,z) - f(t,y',z')|_H^2 \le k \left(|y - y'|_H^2 + |(z - z') \, b(t)|^2_{L_2(H)} \right). \tag{4}$$

Let us now introduce the following spaces.

$$\hat{L}^2_{\mathcal{F}}(0,T;H) := \{\phi : [0,T]\times\Omega \to H, \text{ progressively measurable and such that } \mathbb{E}\left[\int_0^T e^{\gamma\, c_t}|\phi(t)|^2_H\, dc_t\right] < \infty\}.$$

$$\hat{\Lambda}^2(H;\mathcal{P},M) := \{\phi : [0,T]\times\Omega \to L_2(H), \text{ predictable and such that } \mathbb{E}\left[\int_0^T e^{\gamma\, c_t}|\phi(t)\, b(t)|^2_{L_2(H)}\, dc_t\right] < \infty\}.$$

$$\mathcal{S}^2(H) := \{\phi : [0,T]\times\Omega \to H, \text{ cadlag, adapted and such that } \mathbb{E}\left[\sup_{0\le t\le T}(e^{\gamma\, c_t}|\phi(t)|^2_H)\right] < \infty\}.$$

$$\hat{\mathcal{M}}^2_{[0,T]}(H) := \{N \in \mathcal{M}^2_{[0,T]}(H),\ \mathbb{E}\left[\int_0^T e^{\gamma\, c_t}\, d\langle N\rangle_t\right] < \infty\}.$$

$$\mathcal{B}^2(H) := \hat{L}^2_{\mathcal{F}}(0,T;H) \times \hat{\Lambda}^2(H;\mathcal{P},M).$$

Then $\mathcal{B}^2(H)$ is a separable Hilbert space with the norm

$$\|(\phi_1,\phi_2)\|^2_{\mathcal{B}^2(H)} := \mathbb{E}\left[\int_0^T e^{\gamma\, c_t}\,|\phi_1(t)|^2_H\, dc_t\right] + \mathbb{E}\left[\int_0^T e^{\gamma\, c_t}\,|\phi_2(t)\, b(t)|^2_{L_2(H)}\, dc_t\right].$$

A *solution* of (3) is a triple $(Y,Z,N) \in \mathcal{B}^2(H)\times\hat{\mathcal{M}}^2_{[0,T]}(H)$ such that for all $t\in[0,T]$, we have *a.s.*

$$Y(t) = \xi + \int_t^T f(s,Y(s),Z(s))\, dc_s - \int_t^T Z(s)\, dM(s) - \int_t^T dN(s), \quad (5)$$

$N(0) = 0$ and M and N are V.S.O.

The main result is the following theorem.

Theorem 3.1. *Suppose that (H1)–(H5) hold with a parameter γ being large enough. Then there exists a unique solution $(Y,Z,N) \in \mathcal{B}^2(H)\times\hat{\mathcal{M}}^2_{[0,T]}(H)$ of (3).*

Moreover, $Y \in \mathcal{S}^2(H)$.

B. In preparation to proving this theorem we give the following proposition which gives more properties to the integrand X and to the martingale N in the above representation identity (2). Recall the information in Theorem 2.1.

Proposition 3.1. *Under the conditions in Theorem 2.1, if moreover the martingale L belongs to the space $\hat{\mathcal{M}}^2_{[0,T]}(H)$, then the processes X and N in the representation identity (2) lie in the spaces $\hat{\Lambda}^2(H;\mathcal{P},M)$ and $\hat{\mathcal{M}}^2_{[0,T]}(H)$, respectively.*

Proof. Recall first that

$$\mathbb{E}\left[\left|\int_0^T e^{(\gamma/2)\,c_s}\,(X(s)\,dM(s)+dN(s))\right|_H^2\right] = \mathbb{E}\left[\int_0^T e^{\gamma\,c_s}\,|X(s)\,b(s)|^2_{L_2(H)}\,dc_s\right] + \mathbb{E}\left[\int_0^T e^{\gamma\,c_s}\,d\langle N\rangle_s\right].$$

Thus

$$\begin{aligned}
\mathbb{E}&\left[\int_0^T e^{\gamma\,c_s}\,|X(s)\,b(s)|^2_{L_2(H)}\,dc_s\right] + \mathbb{E}\left[\int_0^T e^{\gamma\,c_s}\,d\langle N\rangle_s\right]\\
&= \mathbb{E}\left[\left|\int_0^T e^{(\gamma/2)\,c_s}\,dL(s)\right|_H^2\right]\\
&= \mathbb{E}\left[\int_0^T e^{\gamma\,c_s}\,d\langle L\rangle_s\right]\\
&< \infty.
\end{aligned}$$

This shows that $(X,N)\in\hat{\Lambda}^2(H;\mathcal{P},M)\times\hat{\mathcal{M}}^2_{[0,T]}(H)$. □

Before we can establish the proof of Theorem 3.1 we need to give some lemmas.

Lemma 3.1. *Assume that (H1)–(H5) hold. Let $(y,z)\in\mathcal{B}^2(H)$. The process Y defined by*

$$Y(t) = \mathbb{E}\left[\xi + \int_t^T f(s,y(s),z(s))\,dc_s \;\middle|\; \mathcal{F}_t\right], \quad 0\le t\le T, \tag{6}$$

belongs to the space $\hat{L}^2_{\mathcal{F}}(0,T;H)$, and moreover $Y\in\mathcal{S}^2(H)$.

Proof. Note that it holds

$$\begin{aligned}
&\left(\int_t^T |f(s,y(s),z(s))|_H\,dc_s\right)^2\\
&\quad= \left(\int_t^T e^{(-\beta/2)c_s}\,e^{(\beta/2)c_s}\,|f(s,y(s),z(s))|_H\,dc_s\right)^2\\
&\quad\le \frac{1}{\beta}\,e^{-\beta c_t}\int_t^T e^{\beta c_s}\,|f(s,y(s),z(s))|_H^2\,dc_s\,,
\end{aligned} \tag{7}$$

a.s. for all $t \in [0,T]$ and for all $\beta > 0$. Therefore, putting $\beta := \gamma/2$ in this inequality and using Fubini's theorem give

$$\int_0^T e^{\gamma c_t} \left(\int_t^T |f(s,y(s),z(s))|_H \, dc_s \right)^2 dc_t \leq \frac{4}{\gamma^2} \int_0^T e^{\gamma c_t} \, |f(t,y(t),z(t))|_H^2 \, dc_t, \quad a.s. \tag{8}$$

This together with (4) yields

$$\begin{aligned} \int_0^T e^{\gamma c_t} \left(\int_t^T |f(s,y(s),z(s))|_H \, dc_s \right)^2 dc_t &\leq \frac{8k}{\gamma^2} \int_0^T e^{\gamma c_t} \, |y(t)|_H^2 \, dc_t \\ &+ \frac{8k}{\gamma^2} \int_0^T e^{\gamma c_t} \, |z(t)\, b(t)|_{L_2(H)}^2 \, dc_t \\ &+ \frac{8}{\gamma^2} \int_0^T e^{\gamma c_t} \, |f(t,0,0)|_H^2 \, dc_t, \quad a.s. \end{aligned} \tag{9}$$

Now by applying Cauchy-Schwartz inequality and this inequality, we find that

$$\begin{aligned} &\mathbb{E}\left[\int_0^T e^{\gamma c_t} \, |Y(t)|_H^2 \, dc_t \right] \\ &\leq 2\,\mathbb{E}\left(\int_0^T \mathbb{E}\left[e^{\gamma c_t} \, |\xi|_H^2 \mid \mathcal{F}_t \right] dc_t \right) \\ &\quad + 2\,\mathbb{E}\left(\int_0^T \mathbb{E}\left[e^{\gamma c_t} \left(\int_t^T |f(s,y(s),z(s))|_H \, dc_s \right)^2 \,\Big|\, \mathcal{F}_t \right] dc_t \right) \\ &\leq \frac{2}{\gamma}\,\mathbb{E}\left[e^{\gamma c_T} \, |\xi|_H^2 \right] \\ &\quad + \frac{16k}{\gamma^2}\,\mathbb{E}\left[\int_0^T e^{\gamma c_t} \, |y(t)|_H^2 \, dc_t \right] \\ &\quad + \frac{16k}{\gamma^2}\,\mathbb{E}\left[\int_0^T e^{\gamma c_t} \, |z(t)\, b(t)|_{L_2(H)}^2 \, dc_t \right] \\ &\quad + \frac{16}{\gamma^2}\,\mathbb{E}\left[\int_0^T e^{\gamma c_t} \, |f(t,0,0)|_H^2 \, dc_t \right]. \end{aligned} \tag{10}$$

In particular, the conditions (H4) and (H5) and this result show that $Y \in \hat{L}^2_{\mathcal{F}}(0,T;H)$.

It remains to prove that $Y \in \mathcal{S}^2(H)$. Applying (6), (7) (with $\beta = \gamma$), Doob's inequality (see e.g. [13]) and the assumptions (4), (H4) and (H5) shows that

$$\begin{aligned}
&\mathbb{E}\Big[\sup_{0\leq t\leq T}(e^{\gamma c_t}\,|Y(t)|_H^2)\Big]\\
&= \mathbb{E}\left[\sup_{0\leq t\leq T}\left(e^{\gamma c_t}\,\left|\mathbb{E}\left[\xi + \int_t^T f(s,y(s),z(s))\,dc_s \,\Big|\, \mathcal{F}_t\right]\right|_H^2\right)\right]\\
&\leq 2\,\mathbb{E}\left[\sup_{0\leq t\leq T}\left(\mathbb{E}\left[e^{\gamma c_T}\,|\xi|_H^2 \right.\right.\right.\\
&\qquad\qquad \left.\left.\left. + e^{\gamma c_t}\left(\int_t^T |f(s,y(s),z(s))|_H\,dc_s\right)^2 \,\Big|\, \mathcal{F}_t\right]\right)\right]\\
&\leq 8\,\mathbb{E}\left[e^{\gamma c_T}\,|\xi|_H^2\right]\\
&\quad + \frac{16k}{\gamma}\,\mathbb{E}\left[\int_0^T e^{\gamma c_s}\,|y(s)|_H^2\,dc_s\right]\\
&\quad + \frac{16k}{\gamma}\,\mathbb{E}\left[\int_0^T e^{\gamma c_s}\,|z(s)\,b(s)|_{L_2(H)}^2\,dc_s\right]\\
&\quad + \frac{16}{\gamma}\,\mathbb{E}\left[\int_0^T e^{\gamma c_s}\,|f(s,0,0)|_H^2\,dc_s\right] < \infty,
\end{aligned} \tag{11}$$

which completes the proof. □

Lemma 3.2. *Under the same conditions in Lemma 3.1 the process*

$$K(t) = \mathbb{E}\left[\xi + \int_0^T f(s,y(s),z(s))\,dc_s \,\Big|\, \mathcal{F}_t\right], \quad 0 \leq t \leq T, \tag{12}$$

lies in $\hat{\mathcal{M}}^2_{[0,T]}(H)$.

Proof. Let (Z,N) be the unique processes in $\hat{\Lambda}^2(H;\mathcal{P},M) \times \mathcal{M}^2_{[0,T]}(H)$ given by applying Theorem 2.1 through the formula:

$$K(t) = K(0) + \int_0^t Z(s)\,dM(s) + N(t), \quad 0 \leq t \leq T, \tag{13}$$

such that $N(0) = 0$ and M and N are V.S.O. Recall that $K(0) = Y(0)$ and $K(t) = Y(t) + \int_0^t f(s,y(s),z(s))\,dc_s$ for each t, which comes from the definition of Y in (6). We can then apply the integration by parts and use this fact to find that

$$\mathbb{E}\left[\int_0^T e^{\gamma c_s}\, d\langle K\rangle_s\right] = \mathbb{E}\left[\left|\xi - Y(0) + \int_0^T f(s, y(s), z(s))\, dc_s\right|_H^2\right] + \gamma\,\mathbb{E}\left[\int_0^T e^{\gamma c_s}\left|\xi - Y(s) + \int_s^T f(r, y(r), z(r))\, dc_r\right|_H^2 dc_s\right]. \quad (14)$$

Thus

$$\begin{aligned}\mathbb{E}\left[\int_0^T e^{\gamma c_s}\, d\langle K\rangle_s\right] &\le 3\left(\mathbb{E}\left[|\xi|_H^2\right] + \mathbb{E}\left[|Y(0)|_H^2\right]\right) + 3\,\mathbb{E}\left(\int_0^T |f(s, y(s), z(s))|_H\, dc_s\right)^2 \\ &\quad + 3\gamma\,\mathbb{E}\left[\int_0^T e^{\gamma c_s}\,|\xi|_H^2\, dc_s\right] + 3\gamma\,\mathbb{E}\left[\int_0^T e^{\gamma c_s}\,|Y(s)|_H^2\, dc_s\right] \\ &\quad + 3\gamma\,\mathbb{E}\left[\int_0^T e^{\gamma c_s}\left(\int_s^T |f(r, y(r), z(r))|_H\, dc_r\right)^2 dc_s\right]. \quad (15)\end{aligned}$$

Note that

$$\begin{aligned}\mathbb{E}\left[|\xi|_H^2 + |Y(0)|_H^2\right] &\le \mathbb{E}\left[e^{\gamma c_T}\,|\xi|_H^2\right] + \mathbb{E}\left[\left(\sup_{0\le t\le T} e^{\gamma c_t}\,|Y(t)|_H^2\right)\right] \\ &\le 9\,\mathbb{E}\left[e^{\gamma c_T}\,|\xi|_H^2\right] + \frac{16k}{\gamma}\,\mathbb{E}\left[\int_0^T e^{\gamma c_s}\,|y(s)|_H^2\, dc_s\right] \\ &\quad + \frac{16k}{\gamma}\,\mathbb{E}\left[\int_0^T e^{\gamma c_s}\,|z(s)\, b(s)|_{L_2(H)}^2\, dc_s\right] \\ &\quad + \frac{16}{\gamma}\,\mathbb{E}\left[\int_0^T e^{\gamma c_s}\,|f(s, 0, 0)|_H^2\, dc_s\right], \quad (16)\end{aligned}$$

by using (11). Now substituting (16) in (15) and applying (7) (with $\beta = \gamma$), (4), (10) and (9) give the following result.

$$\begin{aligned}\mathbb{E}\left[\int_0^T e^{\gamma c_s}\, d\langle K\rangle_s\right] &\le 36\,\mathbb{E}\left[e^{\gamma c_T}\,|\xi|_H^2\right] + \frac{126k}{\gamma}\,\mathbb{E}\left[\int_0^T e^{\gamma c_s}\,|y(s)|_H^2\, dc_s\right] \\ &\quad + \frac{126k}{\gamma}\,\mathbb{E}\left[\int_0^T e^{\gamma c_s}\,|z(s)\, b(s)|_{L_2(H)}^2\, dc_s\right] \\ &\quad + \frac{126}{\gamma}\,\mathbb{E}\left[\int_0^T e^{\gamma c_s}\,|f(s, 0, 0)|_H^2\, dc_s\right]. \quad (17)\end{aligned}$$

Thus by applying the conditions in the lemma we complete the proof. □

C. We are now ready to establish the proof of Theorem 3.1.

Proof of Theorem 3.1. Define the mapping Φ on $\mathcal{B}^2(H)$ by $\Phi(y,z) = (Y,Z)$, such that Y is the right continuous version of the semimartingale given by (6) and Z (and also the martingale N) are the unique processes given by the decomposition (13) as discussed earlier at the beginning of the proof of the preceding lemma. Recall that N is V.S.O. to M and $N(0) = 0$. According to Lemma 3.2 and Proposition 3.1 we conclude that $Z \in \hat{\Lambda}^2(H; \mathcal{P}, M)$ and $N \in \hat{\mathcal{M}}^2_{[0,T]}(H)$. Therefore Φ maps $\mathcal{B}^2(H)$ into itself, and (Y, Z, N) is the solution of the following BSDE:

$$Y(t) = \xi + \int_t^T f(s, y(s), z(s))\, dc_s - \int_t^T Z(s)\, dM(s) - \int_t^T dN(s), \quad (18)$$

where $t \in [0,T]$. The rest of the proof is standard, but we give it here for completeness.

We shall show in the following that Φ is a contraction mapping on $\mathcal{B}^2(H)$. Take two elements (y_1, z_2) and (y_2, z_2) of $\mathcal{B}^2(H)$ and let (Y_1, Z_1) and (Y_2, Z_2) denote respectively their images in $\mathcal{B}^2(H)$ under Φ. Thus (Y_i, Z_i, N_i) is the solution of the BSDE (18) with generator $f(t, y_i(t), z_i(t))$ and the terminal value ξ, for $i = 1, 2$. Denote $\delta Y = Y_1 - Y_2$, $\delta Z = Z_1 - Z_2$ and $\delta N = N_1 - N_2$. It is clear that $(\delta Y, \delta Z) \in \mathcal{B}^2(H)$ and $\delta N \in \hat{\mathcal{M}}^2_{[0,T]}(H)$. We have *a.s.* for all $t \in [0,T]$,

$$\begin{aligned} \delta Y(t) = &\int_t^T \left(f(s, y_1(s), z_1(s)) - f(s, y_2(s), z_2(s)) \right) dc_s \\ &- \int_t^T \delta Z(s)\, dM(s) - \int_t^T d(\delta N)(s). \end{aligned}$$

Since this equation is of the sort of the BSDE (18), we can obtain two estimates similar to those in (10) and (17). In particular, we must have

$$\begin{aligned} &\mathbb{E}\left[\int_0^T e^{\gamma c_t} \, |\delta Y(t)|_H^2 \, dc_t \right] + \mathbb{E}\left[\int_0^T e^{\gamma c_t} \, |\delta z(t)\, b(t)|^2_{L_2(H)} \, dc_t \right] \\ &\quad \le \frac{C}{\gamma} \left(\mathbb{E}\left[\int_0^T e^{\gamma c_t} \, |\delta y(t)|_H^2 \, dc_t \right] + \mathbb{E}\left[\int_0^T e^{\gamma c_t} \, |\delta z(t)\, b(t)|^2_{L_2(H)} \, dc_t \right] \right), \end{aligned} \quad (19)$$

for some positive constant C. Hence, if we choose $\gamma > C$, we find that Φ is a contraction mapping on $\mathcal{B}^2(H)$. Consequently, Φ has a unique fixed point in $\mathcal{B}^2(H)$, call it (Y, Z). Now from the definition of Φ it can be seen that (Y, Z, N) is the unique solution of the BSDE (3), where the martingale $N \in \hat{\mathcal{M}}^2_{[0,T]}(H)$ is given with the help of Theorem 2.1 by

$$\mathbb{E}\left[\xi + \int_0^T f(s, Y(s), Z(s))\, dc_s \,\middle|\, \mathcal{F}_t\right]$$
$$= Y(0) + \int_0^t Z(s)\, dM(s) + N(t), \quad 0 \le t \le T.$$

Finally, this solution Y lies in $\mathcal{S}^2(H)$ as deduced from (11). □

D. Note that the process Y which solves the BSDE (3) is only known to have a right continuous version, so it may develop a jump. In [4] we give a condition on the filtration $\{\mathcal{F}_t\}_{t\ge0}$ to guarantee the continuity of the martingale N and hence Y.

We now close the paper by the following remark. Assume for simplicity that the space H is the real space $\mathbb{R}$. Assume also that M is the martingale given by the formula $M(t) = \int_0^t f(s)\, dB(s)$, $t \ge 0$, where $f \in L^2([0,T];\mathbb{R})$ (or even random) and B is a Brownian motion taking its values in $\mathbb{R}$. If $f(s) > 0$ for each $s \ge 0$, then we find that $\mathcal{F}_t(M) = \mathcal{F}_t(B)$ for each $t \ge 0$. Therefore by making use of the unique representation of martingales in Theorem 2.1 and the Brownian martingale representation theorem (see [13, Theorem 3.4, p. 200]), one concludes that the martingale N in Theorem 2.1 vanishes almost surely. This tells in particular that the BSDE (3) becomes similar to the BSDEs studied by Pardoux and Peng in [12] but with the variable Zf replacing Z there; see the BSDE (1).

References

1. A. Al-Hussein, *Backward stochastic evolution equations in infinite dimensions*, Ph.D. thesis, Warwick University, UK, 2002.
2. A. Al-Hussein, "Backward stochastic differential equations in infinite dimensions and applications", *Arab J. Math. Sc.* **10**, no. 2, 1–42 (2004).
3. A. Al-Hussein, "Backward stochastic evolution equations", preprint (submitted).
4. A. Al-Hussein, "Backward stochastic partial differential equations in infinite dimensions", *Random Oper. and Stoch. Equ.* **14**, no. 1, 1–22 (2006).
5. Jean-Michel Bismut, "Théorie probabiliste du contrôle des diffusions". *Mem. Amer. Math. Soc.* **4**, no. 167 (1976).
6. Philippe Briand, Bernard Delyon, Jean Mémin, "On the robustness of backward stochastic differential equations", *Stochastic Process. Appl.* **97**, no. 2, 229–253 (2002).

7. N. El Karoui, S.-J. Huang, "A general result of existence and uniqueness of backward stochastic differential equations", *Backward stochastic differential equations (Paris, 1995–1996)*, 27–36, *Pitman Res. Notes Math. Ser.*, 364, Longman, Harlow, 1997.
8. N. El Karoui, E. Pardoux, M. C. Quenez, "Reflected backward SDEs and American options", *Numerical methods in finance*, 215–231, Publ. Newton Inst., Cambridge Univ. Press, Cambridge, 1997.
9. Michel Métivier, *Semimartingales. A course on stochastic processes.* de Gruyter Studies in Mathematics, 2. Walter de Gruyter & Co., Berlin-New York, 1982.
10. Michel Métivier, Jean Pellaumail, *Stochastic Integration. Probability and Mathematical Statistics*, Academic Press, Harcourt Brace Jovanovich, Publishers, New York-London-Toronto, Ont., 1980.
11. Jean-Yves Ouvrard, "Représentation de martingales vectorielles de carré intégrable à valeurs dans des espaces de Hilbert réels séparables" (French), *Z. Wahrscheinlichkeitstheorie und Verw. Gebiete* **33**, no. 3, 195–208 (1975/76).
12. É. Pardoux, S. G. Peng, "Adapted solution of a backward stochastic differential equation", *Systems Control Lett.* **14**, no. 1, 55–61 (1990).
13. Daniel Revuz, Marc Yor, *Continuous martingales and Brownian motion*, third edition, *Grundlehren der Mathematischen Wissenschaften* [Fundamental Principles of Mathematical Sciences], 293, Springer-Verlag, Berlin, 1999.
14. B. L. Rozovskiĭ, *Stochastic evolution systems. Linear theory and applications to nonlinear filtering*, translated from the Russian by A. Yarkho, *Mathematics and its Applications (Soviet Series)*, 35, Kluwer Academic Publishers Group, Dordrecht, 1990.

AbdulRahman Al-Hussein
Department of Mathematics, College of Science, Al-Qassim University,
P. O. Box 237, Buraidah 81999, Saudi Arabia
alhusseinqu@hotmail.com, alhusein@ksu.edu.sa

PARTIAL UNITARITY ARISING FROM QUADRATIC QUANTUM WHITE NOISE

Wided Ayed (*Inst. Prép. aux Études d'Ingénieures, El Merezka*),
Nobuaki Obata (*Tohoku University, Sendai*),
Habib Ouerdiane (*Université de Tunis El Manar*)

In general, the solution to a normal-ordered white noise differential equation involving quadratic quantum white noise is a white noise operator and is not an operator acting in the L^2-space over the original Gaussian space where the quantum white noise is defined. The solution happens to be a unitary operator on a certain subspace of the L^2-space over a Gaussian space with different variance. This regularity property is referred to as *partial unitarity*.

1. Introduction

Given a quantum stochastic process $\{L_t\}$, we consider a normal-ordered white noise differential equation

$$\frac{d\Xi_t}{dt} = L_t \diamond \Xi_t, \qquad \Xi\big|_{t=0} = \Xi_0, \tag{1}$$

where $\diamond$ is the Wick product (or normal-ordered product). Roughly speaking, the unique solution is always found in a space of white noise operators, suitably chosen according to the coefficient $\{L_t\}$ and the initial value Ξ_0, see e.g., Chung–Ji–Obata [4] and Ji–Obata [6]. Let $\{a_t, a_t^*\}$ be the quantum white noise. If L_t is a linear combination of $\{a_t^* a_t, a_t, a_t^*, 1\}$, the equation (1) is reduced essentially to a usual quantum stochastic differential equation for which the quantum Itô theory works well, see Parthasarathy [15]. As is well known, the higher powers of quantum white noise have rather singular nature but are well formulated in quantum white noise theory. The case when $\{L_t\}$ involves a quadratic quantum white noise $\{a_t^2, a_t^{*2}\}$ is a non-trivial step going beyond the traditional quantum Itô theory and the regularity properties of the solution are of great interest. Recall also that the quadratic quantum white noise is related to the Lévy Laplacian, see Ji–Obata–Ouerdiane [9] and Obata [14].

This paper is devoted to one of the simplest cases. We consider

$$\frac{d\Xi}{dt} = \gamma\, a_t^2 \diamond \Xi, \qquad \Xi\big|_{t=0} = \mathcal{H}_{a,b}\,, \tag{2}$$

where $\gamma, a, b \in \mathbb{C}$ are constant numbers and $\mathcal{H}_{a,b}$ a Fourier–Gauss transform. In general, the solution is merely a white noise operator. We shall prove that the solution happens to be unitary on a certain subspace of L^2-space over a Gaussian space whose variance is different from the one of the original space where the quantum white noise is defined. This property is called *partial unitarity*. Our result is relevant to unitarity of a (generalized) Fourier–Gauss transform investigated by Ji–Obata [7, 8]. The main results will be stated in Section 5.

There are different approaches to the quadratic quantum white noise, see e.g., Accardi–Amosov–Franz [1], Accardi–Franz–Skeide [2], Lytvynov [12], and references cited therein.

2. Generalized Fourier–Gauss transforms

We adopt mostly the same notations as in [7]. Let us start with a real Gelfand triple

$$\mathcal{N} = \mathcal{S}(\mathbb{R}) \subset H = L^2(\mathbb{R}, dt) \subset \mathcal{N}^* = \mathcal{S}'(\mathbb{R}), \tag{3}$$

where $H = L^2(\mathbb{R})$ is the Hilbert space of $\mathbb{R}$-valued square-integrable functions on the real line $\mathbb{R}$ with respect to the Lebesgue measure dt, $\mathcal{S}(\mathbb{R})$ the space of rapidly decreasing functions and $\mathcal{S}'(\mathbb{R})$ the space of tempered distributions. The canonical bilinear form on $\mathcal{N}^* \times \mathcal{N}$ is denoted by $\langle \cdot, \cdot \rangle$, which is compatible with the inner product of H. By the same symbol we denote the canonical $\mathbb{C}$-bilinear form on $\mathcal{N}_{\mathbb{C}}^* \times \mathcal{N}_{\mathbb{C}}$, where the suffix means the complexification.

With $a \in \mathbb{C}$ and $\xi \in \mathcal{N}_{\mathbb{C}}$ we associate a continuous function $\phi_{a,\xi}$ on $\mathcal{N}^*$ defined by

$$\phi_{a,\xi}(x) = e^{\langle x, \xi \rangle - a\langle \xi, \xi \rangle /2}, \quad x \in \mathcal{N}^*, \tag{4}$$

which we call a *coherent vector* or an *exponential vector*. Let $\mathcal{E}$ be the linear space spanned by $\{\phi_{a,\xi} \,;\, \xi \in \mathcal{N}_{\mathbb{C}}\}$. Due to the obvious relation

$$\phi_{a,\xi} = e^{(a'-a)\langle \xi, \xi \rangle /2} \phi_{a',\xi}, \qquad a, a' \in \mathbb{C}, \quad \xi \in \mathcal{N}_{\mathbb{C}}, \tag{5}$$

the space $\mathcal{E}$ does not depend on the choice of $a \in \mathbb{C}$. In general, two locally convex spaces $\mathcal{X}, \mathcal{Y}$ we denote by $\mathcal{L}(\mathcal{X}, \mathcal{Y})$ the space of continuous operators equipped with the bounded convergence topology. In the next, we will use such space for X and Y are equal to $\mathcal{N}_{\mathbb{C}}$ or $\mathcal{W}$ or their dual spaces.

With a pair $A \in \mathcal{L}(\mathcal{N}_{\mathbb{C}}, \mathcal{N}_{\mathbb{C}}^*)$ and $B \in \mathcal{L}(\mathcal{N}_{\mathbb{C}}, \mathcal{N}_{\mathbb{C}})$ we associate an operator $\mathcal{G}(A, B)$ on $\mathcal{E}$ defined by

$$\mathcal{G}(A, B)\phi_{1,\xi} = e^{\langle A\xi,\xi\rangle/2}\phi_{1,B\xi}\,, \qquad \xi \in \mathcal{N}_{\mathbb{C}}\,.$$

The above formula is sufficient to define a linear operator on $\mathcal{E}$ since the exponential vectors $\{\phi_{a,\xi}\,;\,\xi \in \mathcal{N}_{\mathbb{C}}\}$ are linearly independent. The operator $\mathcal{G}(A, B)$ is called a *generalized Fourier–Gauss transform*. Our definition is due to Chung–Ji [3], while an equivalent definition is given by Lee–Liu [11] in terms of an integral formula.

Lemma 2.1.

(1) $\mathcal{G}(A_1, B_1)\,\mathcal{G}(A_2, B_2) = \mathcal{G}(B_2^* A_1 B_2 + A_2, B_1 B_2)$.

(2) $\mathcal{G}(A, B) = 1$ *(the identity operator on* $\mathcal{E}$*) if and only if* $A = 0$ *and* $B = 1$ *(the identity operator on* $\mathcal{N}_{\mathbb{C}}$*).*

(3) $\mathcal{G}(A, B)$ *is invertible if and only if so is* B*, i.e.,* $B \in GL(\mathcal{N}_{\mathbb{C}})$*. In that case,* $\mathcal{G}(A, B)^{-1} = \mathcal{G}(-(B^{-1})^* A B^{-1}, B^{-1})$.

The proof is immediate from definition. In particular,

$$\begin{aligned}\mathfrak{G} &\equiv \{\mathcal{G}(A, B)\,;\, A \in \mathcal{L}(\mathcal{N}_{\mathbb{C}}, \mathcal{N}_{\mathbb{C}}^*),\ B \in GL(\mathcal{N}_{\mathbb{C}})\}\\ &\cong \mathcal{L}(\mathcal{N}_{\mathbb{C}}, \mathcal{N}_{\mathbb{C}}^*) \rtimes GL(\mathcal{N}_{\mathbb{C}})\end{aligned}$$

becomes a group of linear automorphisms of $\mathcal{E}$. If both A, B are scalar operators, say, $A = \alpha 1$ and $B = \beta 1$, we write simply $\mathcal{G}(\alpha, \beta)$ and is called a *Fourier–Gauss transform*. We have

$$\mathcal{G}(\alpha, \beta)\phi_{1,\xi} = e^{\alpha\langle\xi,\xi\rangle/2}\phi_{1,\beta\xi}\,, \qquad \xi \in \mathcal{N}_{\mathbb{C}}\,. \tag{6}$$

We naturally come to a subgroup of $\mathfrak{G}$:

$$\mathfrak{G}_0 = \{\mathcal{G}(\alpha, \beta)\,;\, \alpha \in \mathbb{C},\ \beta \in \mathbb{C}^\times\} \cong \mathbb{C} \rtimes \mathbb{C}^\times,$$

where $\mathbb{C}^\times$ is the multiplicative group of non-zero complex numbers.

For later use we define one-parameter subgroups of $\mathfrak{G}_0$. First, for $a \in \mathbb{C}$ we define

$$T_a = \mathcal{G}(a, 1)\ :\ \phi_{1,\xi} \mapsto e^{a\langle\xi,\xi\rangle/2}\phi_{1,\xi}\,, \qquad \xi \in \mathcal{N}_{\mathbb{C}}.$$

It follows immediately that

$$T_a T_{a'} = T_{a+a'}\,, \qquad T_a^{-1} = T_{-a}\,.$$

Moreover, by a straightforward computation we obtain

$$T_{a-1+2b} : \phi_{a,\xi} \mapsto e^{b\langle\xi,\xi\rangle}\phi_{1,\xi}\,, \qquad \xi \in \mathcal{N}_{\mathbb{C}}\,, \quad a, b \in \mathbb{C}. \tag{7}$$

Next, for $a, b \in \mathbb{C}$ let $\mathcal{H}_{a,b}$ be a linear operator on $\mathcal{E}$ by

$$\mathcal{H}_{a,b} : \phi_{a,\xi} \mapsto \phi_{a,b\xi}, \qquad \xi \in \mathcal{N}_{\mathbb{C}}. \tag{8}$$

Obviously,

$$\mathcal{H}_{a,b}\mathcal{H}_{a,b'} = \mathcal{H}_{a,bb'}, \qquad \mathcal{H}_{a,b}^{-1} = \mathcal{H}_{a,b^{-1}} \text{ for } b \neq 0.$$

On the other hand, by straightforward computation we obtain

$$\mathcal{H}_{a,b}\phi_{1,\xi} = e^{(a-1)(1-b^2)\langle\xi,\xi\rangle/2}\,\phi_{1,b\xi}\,,$$

which reads

$$\mathcal{H}_{a,b} = \mathcal{G}((a-1)(1-b^2), b). \tag{9}$$

3. Unitarity

The *Gaussian measure* with variance $a > 0$ is a probability measure μ_a on $\mathcal{N}^*$ uniquely specified by

$$e^{-a\langle\xi,\xi\rangle/2} = \int_{\mathcal{N}^*} e^{i\langle x,\xi\rangle}\mu_a(dx), \qquad \xi \in \mathcal{N}.$$

Then we have

$$\langle\!\langle \phi_{a,\xi}, \phi_{a,\eta} \rangle\!\rangle_{\mu_a} \equiv \int_{\mathcal{N}^*} \phi_{a,\xi}(x)\phi_{a,\eta}(x)\,\mu_a(dx) = e^{a\langle\xi,\eta\rangle}, \quad \xi, \eta \in \mathcal{N}_{\mathbb{C}}\,. \tag{10}$$

Lemma 3.1. *For $a > 0$ and $|b| = 1$, the linear automorphism $\mathcal{H}_{a,b}$ of $\mathcal{E}$ extends uniquely to a unitary operator $\tilde{\mathcal{H}}_{a,b}$ on $L^2(\mathcal{N}^*, \mu_a)$.*

Proof. Note that the inner product of $L^2(\mathcal{N}^*, \mu_a)$ is defined by $\langle\!\langle \bar{f}, g \rangle\!\rangle$. Since $\mathcal{E} \subset L^2(\mathcal{N}^*, \mu_a)$ is a dense subspace, it is sufficient to show that

$$\langle\!\langle \overline{\mathcal{H}_{a,b}f}, \mathcal{H}_{a,b}g \rangle\!\rangle_{\mu_a} = \langle\!\langle \overline{f}, g \rangle\!\rangle_{\mu_a}\,, \qquad f, g \in \mathcal{E}.$$

Verification of the above identity is straightforward from (10). □

Let $I \subset \mathbb{R}$ be a closed (finite or infinite) interval. We denote by $\mathcal{E}_I$ the subspace of $\mathcal{E}$ spanned by $\{\phi_{a,\xi}\,;\, \xi \in \mathcal{N}_{\mathbb{C}},\ \mathrm{supp}\,\xi \subset I\}$. By (5), $\mathcal{E}_I$ does not depend on the choice of $a \in \mathbb{C}$ either. In view of the action (6), we are ready to claim the following

Lemma 3.2. *Each* $\mathcal{G}(\alpha,\beta) \in \mathfrak{G}_0$ *induces a linear automorphism of* $\mathcal{E}_I$. *In particular, so is* $\mathcal{H}_{a,b}$ *for any pair* $a, b \in \mathbb{C}$ *with* $b \neq 0$.

For an interval I let 1_I denote the indicator function. The associated multiplication operator is denoted by the same symbol. For $a > 0$ we define a linear map E_I^a from $\mathcal{E}$ into $L^2(\mathcal{N}^*, \mu_a)$ by

$$E_I^a : \phi_{a,\xi} \mapsto \phi_{a,1_I\xi}, \qquad \xi \in \mathcal{N}_{\mathbb{C}}.$$

It is shown that E_I^a extends to a projection on $L^2(\mathcal{N}^*, \mu_a)$, which is denoted by the same symbol. The image of this projection will be denoted by $L^2(\mu_a|\, I)$. It is noted that $\mathcal{E}_I$ is a dense subspace of $L^2(\mu_a|\, I)$.

Now we may state a generalization of Lemma 3.1, the proof of which is similar. We only need to note that $\tilde{\mathcal{H}}_{a,b}$ commutes with the projection E_I^a.

Lemma 3.3. *Let* $I \subset \mathbb{R}$ *be a closed interval and* $a, b \in \mathbb{C}$ *a pair of complex numbers with* $a > 0$ *and* $|b| = 1$. *Then the linear automorphism* $\mathcal{H}_{a,b} \restriction \mathcal{E}_I$ *extends uniquely to a unitary operator on* $L^2(\mu_a|I)$, *which coincides with* $\tilde{\mathcal{H}}_{a,b} \restriction L^2(\mu_a|I)$.

4. White Noise Operators

We take a white noise triple

$$\mathcal{W} \subset \Gamma(H_{\mathbb{C}}) \cong L^2(\mathcal{N}^*, \mu_1) \subset \mathcal{W}^* \tag{11}$$

constructed in the standard manner [5, 6, 10, 13]. Recall that $\Gamma(H_{\mathbb{C}})$ is the Boson Fock space over $H_{\mathbb{C}}$ which is canonically identified with $L^2(\mathcal{N}^*, \mu_1)$ through the Wiener–Itô–Segal isomorphism. For instance, we may take the Hida–Kubo–Takenaka space for (11). The canonical $\mathbb{C}$-bilinear form on $\mathcal{W}^* \times \mathcal{W}$ is denoted by $\langle\!\langle \cdot, \cdot \rangle\!\rangle$.

In general, a continuous operator from $\mathcal{W}$ into $\mathcal{W}^*$ is called a *white noise operator*. Since the canonical injection $\mathcal{W} \to \mathcal{W}^*$ is continuous, we have a natural inclusion $\mathcal{L}(\mathcal{W}, \mathcal{W}) \subset \mathcal{L}(\mathcal{W}, \mathcal{W}^*)$. By simple application of the famous characterization of operator symbols [6, 13] we see that every generalized Fourier–Gauss transform $\mathcal{G}(A, B)$ extends uniquely to a white noise operator in $\mathcal{L}(\mathcal{W}, \mathcal{W})$. In fact, the symbol is given by

$$\langle\!\langle \mathcal{G}(A,B)\phi_{1,\xi}, \phi_{1,\eta} \rangle\!\rangle = \exp\left\{ \frac{1}{2}\langle A\xi, \xi\rangle + \langle B\xi, \eta\rangle \right\}, \qquad \xi, \eta \in \mathcal{N}_{\mathbb{C}},$$

so the check is straightforward. The continuous extension is also called a generalized Fourier–Gauss transform and is denoted by the same symbol. Moreover, we note the following

Proposition 4.1. *Every $\mathcal{G}(A, B) \in \mathfrak{G}$ is a topological linear automorphism of $\mathcal{W}$. In this sense $\mathfrak{G}$ is a subgroup of $GL(\mathcal{W})$.*

We say that $\{L_t\,;\, t \in \mathbb{R}\}$ is a *quantum stochastic process* if $t \mapsto L_t \in \mathcal{L}(\mathcal{W}, \mathcal{W}^*)$ is continuous. Let a_t and a_t^* be the annihilation and creation operators at a time point $t \in \mathbb{R}$, respectively. It is known that both

$$t \mapsto a_t \in \mathcal{L}(\mathcal{W}, \mathcal{W}), \qquad t \mapsto a_t^* \in \mathcal{L}(\mathcal{W}^*, \mathcal{W}^*),$$

are C^∞-maps [6]. The pair $\{a_t, a_t^*\,;\, t \in \mathbb{R}\}$ is called the *quantum white noise process.* We then see that higher powers of quantum white noise (in normal-order) are well defined white noise operators.

As is mentioned in Introduction, we focus on the normal-ordered white noise differential equation:

$$\frac{d\Xi}{dt} = \gamma\, a_t^2 \diamond \Xi, \qquad \Xi\big|_{t=0} = \mathcal{H}_{a,b}\,, \tag{12}$$

where $\gamma, a, b \in \mathbb{C}$ are constant numbers and $\mathcal{H}_{a,b}$ is defined in (8). Recall that $\mathcal{H}_{a,b}$ is a Fourier–Gauss transform and hence, is a white noise operator. By the general theory [4,6] there exists a unique solution to (12) in a space of white noise operators suitably chosen and is given by

$$\Xi_t = \mathcal{H}_{a,b} \diamond \exp \int_0^t \gamma\, a_s^2\, ds = \mathcal{H}_{a,b} \circ \exp \int_0^t \gamma\, a_s^2\, ds, \quad t \geq 0. \tag{13}$$

Here the Wick product $\diamond$ is replaced with the usual product (composition) of operators since the integral contains only annihilation operators.

5. The Main Results

Theorem 5.1. *Let $a, b, \gamma \in \mathbb{C}$ satisfy the following conditions:*

$$|b| = 1, \qquad b \neq \pm 1, \qquad a' \equiv a + \frac{2\gamma}{1-b^2} > 0.$$

Let $\{\Xi_t\}$ be the solution to (12), i.e., given as in (13). Then, for any $t > 0$, the white noise operator Ξ_t possesses the following properties:

(1) *$\Xi_t \restriction \mathcal{E}_{[0,t]}$ extends uniquely to a unitary operator on $L^2(\mu_{a'}|\,[0,t])$.*
(2) *If $a > 0$ in addition, $\Xi_t \restriction \mathcal{E}_{(-\infty,0]}$ and $\Xi_t \restriction \mathcal{E}_{[t,+\infty)}$ extend uniquely to unitary operators on $L^2(\mu_a|\,(-\infty,0])$ and $L^2(\mu_a|\,[t,+\infty))$, respectively.*

As a matter of fact, it will be seen that

$$\Xi_t = \begin{cases} \mathcal{H}_{a',b} & \text{on } \mathcal{E}_{[0,t]} \\ \mathcal{H}_{a,b} & \text{on } \mathcal{E}_{(-\infty,0]} \cup \mathcal{E}_{[t,+\infty)}. \end{cases}$$

Taking into account the canonical factorizations:

$$L^2(\mathcal{N}', \mu_a) = L^2(\mu_a|(-\infty,0]) \otimes L^2(\mu_a|\,[0,t]) \otimes L^2(\mu_a|\,[t,+\infty)),$$
$$L^2(\mathcal{N}', \mu_{a'}) = L^2(\mu_{a'}|(-\infty,0]) \otimes L^2(\mu_{a'}|\,[0,t]) \otimes L^2(\mu_{a'}|\,[t,+\infty)),$$

we see that $\mathcal{E}_{(-\infty,0]} \otimes \mathcal{E}_{[0,t]} \otimes \mathcal{E}_{[t,+\infty)}$ becomes their common dense subspace. Then Theorem 5.1 says that, according to this factorization, we have

$$\Xi_t = \mathcal{H}_{a,b} \otimes \mathcal{H}_{a',b} \otimes \mathcal{H}_{a,b}$$

and each factor in the right hand side extends to a unitary operator on the corresponding subspace of $L^2(\mathcal{N}', \mu_a)$ or $L^2(\mathcal{N}', \mu_{a'})$. We call this property of Ξ_t the *partial unitarity*.

In fact, we prove the following more general result.

Theorem 5.2. *Given $a, b, \gamma \in \mathbb{C}$, let Ξ_t be defined as in* (13). *Assume $|b| = 1$, $b \neq \pm 1$ and choose $a', b' \in \mathbb{C}$ in such a way that*

$$\frac{1}{2}(a - a' + b')(1 - b^2) + \gamma = 0. \tag{14}$$

(1) *If $a' > 0$, the restriction $T_{b'}^{-1}\Xi_t T_{b'} \restriction \mathcal{E}_{[0,t]}$ extends uniquely to a unitary operator on $L^2(\mu_{a'}|\,[0,t])$.*

(2) *If*
$$a'' \equiv a' - \frac{2\gamma}{1-b^2} > 0, \tag{15}$$
then the restrictions $T_{b'}^{-1}\Xi_t T_{b'} \restriction \mathcal{E}_{(-\infty,0]}$ and $T_{b'}^{-1}\Xi_t T_{b'} \restriction \mathcal{E}_{[t,+\infty)}$ extend uniquely to unitary operators on $L^2(\mu_{a''}|\,(-\infty,0])$ and $L^2(\mu_{a''}|\,[t,+\infty))$, respectively.

Theorem 5.1 follows immediately from Theorem 5.2 by setting $b' = 0$. The proof of Theorem 5.2 will be divided into a few steps. The *Gross Laplacian process* is defined by

$$G_t = \int_0^t a_s^2\, ds, \qquad t \geq 0. \tag{16}$$

In fact, $t \mapsto G_t \in \mathcal{L}(\mathcal{W}, \mathcal{W})$ is a C^∞-map.

Lemma 5.1. *For any $\gamma \in \mathbb{C}$ we have*

$$\exp \gamma G_t = \mathcal{G}(2\gamma 1_{[0,t]}, 1). \tag{17}$$

Proof. Since $a_t \phi_{1,\xi} = \xi(t)\phi_{1,\xi}$, we have

$$(\exp \gamma G_t)\phi_{1,\xi} = \exp\left(\gamma \int_0^t \xi(s)^2 ds\right)\phi_{1,\xi} = e^{\langle \gamma 1_{[0,t]}\xi, \xi\rangle}\phi_{1,\xi}\,,$$

which implies (17). □

Lemma 5.2. *Given $a, b, \gamma \in \mathbb{C}$, let Ξ_t be defined as in (13). For any $a', b' \in \mathbb{C}$ and $\xi \in \mathcal{N}_{\mathbb{C}}$ we have*

$$\begin{aligned} &T_{b'}^{-1}\Xi_t T_{b'}\phi_{a',\xi} \\ &= \exp\left\{\frac{1}{2}(a - a' + b')(1 - b^2)\langle \xi, \xi\rangle + \gamma \int_0^t \xi(s)^2 ds\right\}\phi_{a',b\xi}\,. \end{aligned} \tag{18}$$

Proof. Combining (9) and (17), we obtain the solution (13) written in terms of generalized Fourier–Gauss transforms:

$$\Xi_t = \mathcal{H}_{a,b} \circ \exp \gamma G_t = \mathcal{G}((a-1)(1-b^2), b)\,\mathcal{G}(2\gamma 1_{[0,t]}, 1).$$

Then, using $T_{b'} = \mathcal{G}(b', 1)$ and applying the composition rule (Lemma 2.1), we have

$$\begin{aligned} T_{b'}^{-1}\Xi_t T_{b'} &= \mathcal{G}(-b', 1)\,\mathcal{G}((a-1)(1-b^2), b)\,\mathcal{G}(2\gamma 1_{[0,t]}, 1)\,\mathcal{G}(b', 1) \\ &= \mathcal{G}(-b'b^2 + (a-1)(1-b^2), b)\,\mathcal{G}(2\gamma 1_{[0,t]} + b', 1) \\ &= \mathcal{G}(-b'b^2 + (a-1)(1-b^2) + 2\gamma 1_{[0,t]} + b', b) \\ &= \mathcal{G}((a-1+b')(1-b^2) + 2\gamma 1_{[0,t]}, b). \end{aligned}$$

We take the action on $\phi_{a',\xi} = e^{(1-a')\langle\xi,\xi\rangle/2}\phi_{1,\xi}$ to obtain

$$\begin{aligned} &T_{b'}^{-1}\Xi_t T_{b'}\phi_{a',\xi} \\ &= e^{(1-a')\langle\xi,\xi\rangle/2} e^{\langle\{(a-1+b')(1-b^2)+2\gamma 1_{[0,t]}\}\xi,\xi\rangle/2}\phi_{1,b\xi} \\ &= e^{(1-a')\langle\xi,\xi\rangle/2} e^{\langle\{(a-1+b')(1-b^2)+2\gamma 1_{[0,t]}\}\xi,\xi\rangle/2} e^{(a'-1)b^2\langle\xi,\xi\rangle/2}\phi_{a',b\xi} \\ &= e^{\langle\{(a-a'+b')(1-b^2)+2\gamma 1_{[0,t]}\}\xi,\xi\rangle/2}\phi_{a',b\xi}\,, \end{aligned}$$

from which (18) follows immediately. □

Lemma 5.3. *Given* $a, b, \gamma \in \mathbb{C}$, *let* Ξ_t *be defined as in* (13). *Assume* $b \neq \pm 1$ *and choose* $a', b' \in \mathbb{C}$ *in such a way that*

$$\frac{1}{2}(a - a' + b')(1 - b^2) + \gamma = 0. \tag{19}$$

Then, for $t > 0$ *we have*

$$T_{b'}^{-1} \Xi_t T_{b'} = \begin{cases} \mathcal{H}_{a',b} & on\ \mathcal{E}_{[0,t]}\,, \\ \mathcal{H}_{a'',b} & on\ \mathcal{E}_{(-\infty,0]} \cup \mathcal{E}_{[t,+\infty)}\,, \end{cases} \tag{20}$$

where

$$a'' = a' - \frac{2\gamma}{1 - b^2}\,. \tag{21}$$

Proof. Let $\xi \in \mathcal{N}_{\mathbb{C}}$ with $\operatorname{supp} \xi \subset [0, t]$. Then, by (18) and (19) we see that

$$\begin{aligned} T_{b'}^{-1} \Xi_t T_{b'} \phi_{a',\xi} &= \exp\left\{ \frac{1}{2}(a - a' + b')(1 - b^2)\langle \xi, \xi \rangle + \gamma \langle \xi, \xi \rangle \right\} \phi_{a',b\xi} \\ &= \phi_{a',b\xi}\,. \end{aligned}$$

Hence $T_{b'}^{-1} \Xi_t T_{b'} \phi_{a',\xi} = \mathcal{H}_{a',b} \phi_{a',\xi}$ and the first part of (20) is proved.

We next take $\xi \in \mathcal{N}_{\mathbb{C}}$ with $\operatorname{supp} \xi \subset (-\infty, 0] \cup [t, +\infty)$. Again, in view of (18) and (19) we see that

$$T_{b'}^{-1} \Xi_t T_{b'} \phi_{a',\xi} = e^{-\gamma \langle \xi, \xi \rangle} \phi_{a',b\xi}\,,$$

namely,

$$e^{(1-a')\langle \xi, \xi \rangle / 2} T_{b'}^{-1} \Xi_t T_{b'} \phi_{1,\xi} = e^{-\gamma \langle \xi, \xi \rangle + (1-a') b^2 \langle \xi, \xi \rangle / 2} \phi_{1,b\xi}\,.$$

Therefore we have

$$\begin{aligned} T_{b'}^{-1} \Xi_t T_{b'} \phi_{1,\xi} &= e^{-\gamma \langle \xi, \xi \rangle - (1-a')(1-b^2) \langle \xi, \xi \rangle / 2} \phi_{1,b\xi} \\ &= \mathcal{G}(-2\gamma - (1 - a')(1 - b^2), b) \phi_{1,\xi} \\ &= \mathcal{G}\Big(\Big(a' - \frac{2\gamma}{1 - b^2} - 1 \Big)(1 - b^2), b \Big) \phi_{1,\xi}\,. \end{aligned}$$

Taking (9) and (21) into account, we conclude that

$$T_{b'}^{-1} \Xi_t T_{b'} \phi_{1,\xi} = \mathcal{H}_{a'',b} \phi_{1,\xi}\,,$$

which proves the second half of (20). $\square$

Remark 5.1. Lemma 5.3 becomes uninteresting when $b = \pm 1$. In fact, in that case $\gamma = 0$ so that Ξ_t is reduced to a constant independent of t, see (13).

Proof of Theorem 5.2. (1) We already know from Lemma 5.3 that

$$T_{b'}^{-1} \Xi_t T_{b'} \restriction \mathcal{E}_{[0,t]} = \mathcal{H}_{a',b} \restriction \mathcal{E}_{[0,t]}. \tag{22}$$

Noting by assumption that $|b| = 1$ and $a' > 0$, we see from Lemma 3.3 that (22) extends to a unitary operator on $L^2(\mu_{a'} | [0,t])$. The proof of (2) is similar. □

References

1. L. Accardi, G. Amosov, U. Franz, "Second quantized automorphisms of the renormalized square of white noise (RSWN) algebra", *Infin. Dimens. Anal. Quantum Probab. Relat. Top.* **7**, 183–194 (2004).
2. L. Accardi, U. Franz, M. Skeide, "Renormalized squares of white noise and other non-Gaussian noises as Levy processes on real Lie algebras", *Comm. Math. Phys.* **228**, 123–150 (2002).
3. D. M. Chung, U. C. Ji, "Transforms on white noise functionals with their applications to Cauchy problems", *Nagoya Math. J.* **147**, 1–23 (1997).
4. D. M. Chung, U. C. Ji, N. Obata, "Quantum stochastic analysis via white noise operators in weighted Fock space", *Rev. Math. Phys.* **14**, 241–272 (2002).
5. R. Gannoun, R. Hachaichi, H. Ouerdiane, A. Rezgui, "Un théorème de dualité entre espaces de fonctions holomorphes à croissance exponentielle", *J. Funct. Anal.* **171**, 1–14 (2000).
6. U. C. Ji, N. Obata, "Quantum white noise calculus", in *Non-Commutativity, Infinite-Dimensionality and Probability at the Crossroads* (N. Obata, T. Matsui and A. Hora, eds.), World Scientific, 2002, pp. 143–191.
7. U. C. Ji, N. Obata, "Unitarity of Kuo's Fourier–Mehler transform", *Infin. Dimens. Anal. Quantum Probab. Relat. Top.* **7**, 147–154 (2004).
8. U. C. Ji, N. Obata, "Unitarity of generalized Fourier–Gauss transforms", to appear in *Stoch. Anal. Appl.* (2006).
9. U. C. Ji, N. Obata, H. Ouerdiane, "Quantum Lévy Laplacian and associated heat equation", preprint, 2005.
10. H.-H. Kuo, *White Noise Distribution Theory*, CRC Press, 1996.
11. Y.-J. Lee, C.-F. Liu, "A generalization of Mehler transform", in *International Mathematics Conference '94*, World Scientific, 1996, pp. 107–116.
12. E. Lytvynov, "The square of white noise as a Jacobi field", *Infin. Dimens. Anal. Quantum Probab. Relat. Top.* **7**, 619–629 (2004).
13. N. Obata, *White Noise Calculus and Fock Space*, *Lect. Notes in Math.* vol. 1577, Springer-Verlag, 1994.
14. N. Obata, "Quadratic quantum white noises and Lévy Laplacian", *Nonlinear Analysis* **47**, 2437–2448 (2001).

15. K. R. Parthasarathy, *An Introduction to Quantum Stochastic Calculus*, Birkhäuser, 1992.

Wided Ayed
Département de Mathématiques, Institut Préparatoire aux Études d'Ingénieures,
El Merezka, Nabeul, 8000, Tunisia

Nobuaki Obata
Graduate School of Information Sciences,
Tohoku University, Sendai, 980-8579, Japan

Habib Ouerdiane
Département de Mathématiques, Faculté des Sciences,
Université de Tunis El Manar,
Campus Universitaire, Tunis 1060, Tunisia

SCHILDER'S THEOREM FOR GAUSSIAN WHITE NOISE DISTRIBUTIONS

SONIA CHAARI, SOUMAYA GHERYANI,
HABIB OUERDIANE (*University of Tunis El Manar*)

In the theory of large deviations, one of the main results is Schilder's theorem. It gives the large deviation estimates for a family $\{\mu_\epsilon, \epsilon > 0\}$ of measures on some Polish space X, which tends weakly to the Dirac measure δ_p at the point $p \in X$. In this paper, we investigate analogous problems for a family $\{\gamma_\epsilon, \epsilon > 0\}$ of white noise Gaussian measures with mean 0 and variance ϵ on the Schwartz distributions space $S'(\mathbb{R})$. Applications to stochastic differential equations are given.

1. Introduction

In this paper we give an example of large deviation result for a certain family of measures on an infinite dimensional space. For this purpose, let X be a reel nuclear Fréchet space. A function $I : X \longrightarrow [0, +\infty]$ is said to be a good rate function, if it is lower semi-continuous and $\{x \in X, I(x) \leq L\}$ are compact for all $L \geq 0$. We say that a family $\{\gamma_\epsilon, \epsilon > 0\}$ of Borel probability measures on the space X satisfies large deviation principle (LDP) with good rate function I if the following conditions are satisfied

1. (UPPER BOUND) for all closed subsets F in X

$$\limsup_{\epsilon \longrightarrow 0} \epsilon \log(\gamma_\epsilon(F)) \leq - \inf_{y \in F} I(y), \tag{1}$$

2. (LOWER BOUND) for all open sets G in X

$$\liminf_{\epsilon \longrightarrow 0} \epsilon \log(\gamma_\epsilon(G)) \geq - \inf_{y \in G} I(y). \tag{2}$$

Let Λ_γ be the Logarithmic moment generating function, i.e.,

$$\Lambda_\gamma(\xi) := \log\bigg(\int_{S'(\mathbb{R})} e^{\langle y, \xi \rangle} d\gamma(y) \bigg), \quad \xi \in S(\mathbb{R}^d).$$

We denote by Λ_γ^* the Legendre transform of Λ_γ, i.e., for every $\varphi \in S'(\mathbb{R})$,

$$\Lambda_\gamma^*(\varphi) = \sup\bigg\{ \langle \varphi, \xi \rangle - \Lambda_\gamma(\xi); \ \xi \in S(\mathbb{R}) \bigg\}. \tag{3}$$

Then, the rate function is the Legendre transform of the corresponding Logarithmic moment generating function which is also introduced in both Cramer's [3] and Schilder's theorems [16], see also [17]. We remark that, in the one dimensional case where $X = \mathbb{R}$, the rate function I is given by:

$$I(x) = \Lambda_\gamma(x) = \Lambda_\gamma^*(x) = \frac{x^2}{2}.$$

The present paper is organized as follows. In section 2 we recall the structure and concepts of white noise distributions. In section 3 we prove that a family $\{\gamma_\epsilon, \epsilon > 0\}$ of white noise Gaussian measures with mean 0 and variance ϵ on the Schwartz distributions space $S'(\mathbb{R})$ satisfies the large deviation principle with rate function Λ_γ^* given in (9), see lemma 3.1. Section 4 is devoted to applying this large deviation results to the distributions measures associated to the solution of some stochastic differential equations.

2. Notation and preliminaries

Let $N = X + iX$ the complexification of the reel nuclear Fréchet space X and suppose that its topology is defined by a family $\{|.|_p, p \in \mathbb{N}\}$ of increasing Hilbertian norms. We have the representation

$$N = \bigcap_{p\geq 0} N_p = \text{proj} \lim_{p\longrightarrow\infty} N_p$$

where N_p is the completion of N with respect to the norm $|.|_p$. Denote by N_{-p} the topological dual space of the space N_p, then the dual N' of N can be written as

$$N' = \bigcup_{p\geq 0} N_{-p} = \text{ind} \lim_{p\longrightarrow\infty} N_{-p}.$$

Let $\theta : \mathbb{R}_+ \longrightarrow \mathbb{R}_+$ be a Young function, i.e., θ is continuous, convex, strictly increasing and satisfies $\theta(0) = 0$ and $\lim_{x\to\infty} \frac{\theta(x)}{x} = +\infty$. Denote θ^* the Legendre transform of θ: $\theta^*(x) = \sup\{tx - \theta(t); t > 0\}$ for all $x \geq 0$, which also a Young function. Given a complex Banach space $(B, \|.\|)$, let $H(B)$ be the space of entire functions on B, i.e., the space of continuous functions from B to $\mathbb{C}$, whose restriction to all affine lines of B are entire on $\mathbb{C}$. Let $\text{Exp}(B, \theta, m)$ denote the space of all entire functions on B with exponential growth of order θ, and of finite type $m > 0$:

$$\text{Exp}(B, \theta, m) = \Big\{f \in H(B);\ \|f\|_{\theta,m} = \sup_{x\in B} |f(x)|e^{-\theta(m\|x\|)} < +\infty\Big\}.$$

Let also $\|f\|_{\theta,m,p} = \sup_{u \in N_p} |f(x)| e^{-\theta(m|x|_p)}$ for $f \in \mathrm{Exp}(N_p, \theta, m)$. The intersection

$$\mathcal{F}_\theta(N') = \bigcap_{p \geq 0, m > 0} \mathrm{Exp}(N_{-p}, \theta, m),$$

equipped with the projective limit topology, is called the space of entire functions on N' of θ-growth and minimal type. The union

$$\mathcal{G}_\theta(N) = \bigcup_{p \geq 0, m > 0} \mathrm{Exp}(N_p, \theta, m),$$

equipped with the inductive limit topology, is called the space of entire functions on N of θ-growth and (arbitrarily) finite type. Denote by $\mathcal{F}_\theta(N')^*$ the strong dual of the test function space $\mathcal{F}_\theta(N')$. In the sequel we take $N = X + iX$, the complexification of a nuclear Fréchet space X. Let $\mathcal{F}_\theta(N')_+$ denote the cone of positive test functions, i.e., $f \in \mathcal{F}_\theta(N')_+$ if $f(x + i0) \geq 0$ for all y in the topological dual X' of X.

Definition 2.1. The space $\mathcal{F}_\theta(N')^*_+$ of positive distributions is defined as the space of $\phi \in \mathcal{F}_\theta(N')^*$ such that $\langle \phi, f \rangle \geq 0$; $f \in \mathcal{F}_\theta(N')_+$.

We recall the following results on the representation of positive distributions; see [14]:

Theorem 2.1. *Let $\phi \in \mathcal{F}_\theta(N')^*_+$, then there exists a unique Radon measure μ_ϕ on X', such that*

$$\phi(f) = \int_{X'} f(y + i0) d\mu_\phi(y)\,; \;\; f \in \mathcal{F}_\theta(N').$$

Conversely, let μ be a finite, positive Borel measure on X'. Then μ represent a positive distribution in $\mathcal{F}_\theta(N')^*_+$ if and only if μ is supported by some X_{-p}, $p \in \mathbb{N}^*$, and there exists some $m > 0$ such that:

$$\int_{X_{-p}} e^{\theta(m|y|_{-p})}\, d\mu(y) < \infty. \tag{4}$$

We recall also the following estimates given in [15]. For a given $\xi \in X$ and $x \in \mathbb{R}$, let

$$A_{\xi,x} = \{y \in X' : \langle y, \xi \rangle > x\} \tag{5}$$

denote the half-plane in X' associated to ξ and x.

Theorem 2.2. *Let $\phi \in \mathcal{F}_\theta(N')^*_+$ such that ϕ defines a positive Radon measure μ_ϕ on X'. Then for all $\xi \in X$ and $x > 0$, there exists $m > 0$ and $p \in \mathbb{N}$ such that:*

$$\mu_\phi(A_{\xi,x}) \leq \|\widehat{\phi}\|_{\theta,m,p} \exp\left(-\theta\left(\frac{x}{m|\xi|_p}\right)\right), \tag{6}$$

where $\widehat{\phi}$ is the Laplace transform of ϕ.

3. Large deviation for Gaussian measures on $S'(\mathbb{R})$

In this section, we take $X = S(\mathbb{R})$ the Schwartz space of real-valued rapidly decreasing functions on $\mathbb{R}$, and X' the corresponding dual space, i.e., $X' = S'(\mathbb{R})$ the Schwartz distributions space. For every integer n let $H_n(x) = (-1)^n e^{x^2}(\frac{d}{dx})^n e^{-x^2}$ be the Hermite polynomial of degree n and

$$e_n(x) = \frac{1}{\sqrt{\sqrt{\pi}2^n\, n!}} H_n(x) e^{-x^2/2}$$

be the corresponding Hermite function. Then the set $\{e_n; n \geq 0\}$ is an orthonormal basis for the Hilbert space $L^2(\mathbb{R})$. Now for each $p \geq 0$, define

$$|f|_p = \left(\sum_{n=0}^{\infty}(2n+2)^{2p}(f, e_n)^2\right)^{\frac{1}{2}},$$

where $(.,.)$ is the inner product of $L^2(\mathbb{R})$. Let $S_p \equiv \{f \in L^2(\mathbb{R}); |f|_p < \infty\}$. Then we have $S(\mathbb{R}) = \bigcap_{p\geq 0} S_p(\mathbb{R})$ endowed with the projective topology. By the general theory of duality, $S'(\mathbb{R})$ the dual space of $S(\mathbb{R})$ can be written as $S'(\mathbb{R}) = \bigcup_{p\geq 0} S_{-p}(\mathbb{R})$ endowed with the inductive topology, where $S_{-p}(\mathbb{R})$ denotes the topological dual space of $S_p(\mathbb{R})$. Then we have the Gelfand triple

$$S(\mathbb{R}) \hookrightarrow L^2(\mathbb{R}, dx) \hookrightarrow S'(\mathbb{R}). \tag{7}$$

Using the Bochner-Minlos theorem, see [6] and [8], there exists a unique measure γ on $S'(\mathbb{R})$ such that

$$\int_{S'(\mathbb{R})} e^{i\langle y,\xi\rangle}\, d\gamma(y) = e^{-\frac{1}{2}|\xi|_0^2}, \quad \xi \in S(\mathbb{R}),$$

where $\langle .,. \rangle$ denotes the dual paring between $S'(\mathbb{R})$ and $S(\mathbb{R})$ which is realized as an extension of inner product $(.,.)$ on $L^2(\mathbb{R}, dx)$ and $|.|_0$ the corresponding norm.

We begin by introducing the logarithmic moment generating function:

$$\Lambda_\gamma(\xi) := \log\left(\int_{S'(\mathbb{R})} e^{\langle y,\xi\rangle}\, d\gamma(y)\right) = \frac{1}{2}|\xi|_0^2, \quad \forall \xi \in S(\mathbb{R}). \tag{8}$$

The function Λ_γ can be extended to the space $L^2(\mathbb{R})$. In fact, if $\xi \in L^2(\mathbb{R})$ there exists a sequence $(\xi_n)_{n\in\mathbb{N}}$ in $S(\mathbb{R})$ converging to ξ in $L^2(\mathbb{R}, dx)$. Then the sequence $\{\langle .,\xi_n\rangle\}_{n\in\mathbb{N}}$ of random variables is Cauchy in $L^2(S'(\mathbb{R}),\gamma)$. Define

$$\langle .,\xi\rangle := \lim_{n\to\infty}\langle .,\xi_n\rangle, \text{ in } L^2(S'(\mathbb{R}),\gamma).$$

This limit is independent of the choice of the sequence $(\xi_n)_{n\in\mathbb{N}}$, see for example [12]. So for $\xi \in L^2(\mathbb{R})$ the logarithmic moment generating function Λ_γ is defined by

$$\Lambda_\gamma(\xi) = \frac{1}{2}\langle \xi,\xi\rangle = \lim_{n\to\infty}\frac{1}{2}\langle \xi,\xi_n\rangle.$$

Note that Λ_γ is lower semi-continuous and convex function. It is not hard to see that the Legendre transform Λ_γ^* of Λ_γ given in (3) is a lower semi-continuous and convex function, see [17] and [11].

Lemma 3.1. *The Legendre transform of Λ_γ has the following expression:*

$$\Lambda_\gamma^*(\varphi) = \begin{cases} \frac{1}{2}|\varphi|_0^2 = \Lambda_\gamma(\varphi), & \text{if } \varphi \in L^2(\mathbb{R}, dx), \\ +\infty, & \text{if } \varphi \in S'(\mathbb{R})\backslash L^2(\mathbb{R}, dx). \end{cases} \tag{9}$$

Proof. Let $\varphi \in L^2(\mathbb{R}, dx)$, then we have:

$$\begin{aligned} \Lambda_\gamma^*(\varphi) &= \sup\{\langle \varphi,\xi\rangle - \Lambda_\gamma(\xi);\ \xi \in S(\mathbb{R})\} \\ &= \sup\{(\varphi,\xi) - \frac{1}{2}(\xi,\xi);\ \xi \in S(\mathbb{R})\} \\ &= \frac{1}{2}(\varphi,\varphi) = \frac{1}{2}|\varphi|_0^2. \end{aligned}$$

So we obtain

$$\Lambda_\gamma^*(\varphi) = \Lambda_\gamma(\varphi), \text{ for all } \varphi \in L^2(\mathbb{R}, dx).$$

To prove that $\Lambda_\gamma^*(\varphi) = \infty$ for $\varphi \in S'(\mathbb{R})\backslash L^2(\mathbb{R}, dx)$, suppose that $\Lambda_\gamma^*(\varphi) < \infty$ for $\varphi \in S'(\mathbb{R})$. Since the space $L^2(\mathbb{R}, dx)$ is densely embedded in $S'(\mathbb{R})$ with respect to the weak topology, see [5], so for all $\varphi \in S'(\mathbb{R})$ there exists $(\varphi_n)_n \in L^2(\mathbb{R})$ such that $\varphi = \lim_{n\to\infty}\varphi_n$ weakly. By the lower semi-continuity property of the function Λ_γ^*, we have:

$$\Lambda_\gamma^*(\varphi) = \lim_{n\to\infty}\Lambda_\gamma^*(\varphi_n) = \lim_{n\to\infty}\frac{1}{2}|\varphi_n|_0^2 = \frac{1}{2}|\varphi|_0^2 < \infty.$$

Thus, we conclude that $\varphi \in L^2(\mathbb{R}, dx)$ and therefore lemma (3.1) holds. □

For each $\epsilon > 0$, let γ_ϵ denote the image measure of γ under the map

$$M_{\sqrt{\epsilon}} : \xi \mapsto \sqrt{\epsilon}\,\xi, \tag{10}$$

i.e., the Laplace transform is given by

$$\widehat{\gamma}_\epsilon(\varphi) = \widehat{\gamma}(M_{\sqrt{\epsilon}}(\varphi)) = e^{\frac{1}{2}\epsilon|\varphi|_0^2}.$$

Given $\xi \in S_{-p}(\mathbb{R}) \subset S'(\mathbb{R})$, $r > 0$ and $p > 0$, we denote by $B_p(\xi, r)$ the open ball of radius r around a point ξ, and $\overline{B}_p(\xi, r)$ the corresponding closed ball:

$$\overline{B}_p(\xi, r) = \big\{y \in S'(\mathbb{R});\ |y - \xi|_{-p} \le r\big\}.$$

Lemma 3.2. *Let $\xi \in S'(\mathbb{R})$ be given. Then for each $\delta > 0$ there exists $r > 0$ such that*

$$\gamma_\epsilon(\overline{B}_p(\xi, r)) \le \begin{cases} \exp\Big(-\dfrac{\Lambda_\gamma^*(\xi) - \delta}{\epsilon}\Big), & \text{if } \Lambda_\gamma^*(\xi) < \infty, \\ \exp\Big(-\dfrac{1}{\epsilon\delta}\Big), & \text{if } \Lambda_\gamma^*(\xi) = \infty \end{cases} \tag{11}$$

for all $\epsilon > 0$. In particular, if K is a compact subset of $S'(\mathbb{R})$, then

$$\limsup_{\epsilon \to 0} \epsilon \log(\gamma_\epsilon(K)) \le -\inf_K \Lambda_\gamma^*. \tag{12}$$

Proof. First, note that

$$\begin{aligned} \gamma_\epsilon(\overline{B}_p(\xi, r)) &= \gamma\Big(\overline{B}_p\Big(\frac{\xi}{\sqrt{\epsilon}}, \frac{r}{\sqrt{\epsilon}}\Big)\Big) \\ &\le \Big(\sup_{y \in \overline{B}_p\left(\frac{\xi}{\sqrt{\epsilon}}, \frac{r}{\sqrt{\epsilon}}\right)} e^{-\langle y, \frac{\varphi}{\sqrt{\epsilon}}\rangle}\Big) \int_{\overline{B}_p\left(\frac{\xi}{\sqrt{\epsilon}}, \frac{r}{\sqrt{\epsilon}}\right)} e^{\langle y, \frac{\varphi}{\sqrt{\epsilon}}\rangle}\, d\gamma(y) \\ &\qquad\qquad \text{for all } \varphi \in S(\mathbb{R}) \\ &\le \Big(\sup_{y \in \overline{B}_p(\xi, r)} e^{-\frac{1}{\epsilon}\langle y, \varphi\rangle}\Big) \int_{S'(\mathbb{R})} e^{\langle y, \frac{\varphi}{\sqrt{\epsilon}}\rangle}\, d\gamma(y). \end{aligned}$$

For all $y \in S'(\mathbb{R})$, and $\varphi, \xi \in S(\mathbb{R})$, there exists $p \in \mathbb{N}$ such that

$$-|y - \xi|_{-p}\Big|\frac{\varphi}{\epsilon}\Big|_p \le \Big\langle y - \xi, \frac{\varphi}{\epsilon}\Big\rangle \le |y - \xi|_{-p}\Big|\frac{\varphi}{\epsilon}\Big|_p.$$

So for all $y \in \overline{B}_p(\xi, r)$ we have $\langle y, \frac{\varphi}{\epsilon}\rangle \ge \langle \xi, \frac{\varphi}{\epsilon}\rangle - \frac{1}{\epsilon}r|\varphi|_p$. Hence

$$\gamma_\epsilon(\overline{B}_p(\xi, r)) \le \exp\Big(-\frac{1}{\epsilon}\big(\langle \xi, \varphi\rangle - \Lambda_\gamma(\varphi) - r|\varphi|_p\big)\Big).$$

If $\Lambda_\gamma^*(\xi) < \infty$, we choose $\varphi \in S(\mathbb{R})$ such that $\langle \xi, \varphi \rangle - \Lambda_\gamma(\varphi) \geq \Lambda_\gamma^*(\xi) - \frac{\delta}{2}$ and $r = \frac{\delta}{2(1+|\varphi|_p)}$. If $\Lambda_\gamma^*(\xi) = \infty$, choose $\varphi \in S(\mathbb{R})$ such that $\langle \xi, \varphi \rangle - \Lambda_\gamma(\varphi) \geq 1 + \frac{1}{\delta}$ and $r = \frac{1}{(1+|\varphi|_p)}$. So we have

$$\gamma_\epsilon(\overline{B}_p(\xi, r)) \leq \begin{cases} \exp\Big(-\dfrac{\Lambda_\gamma^*(\xi) - \delta}{\epsilon}\Big), & \text{if } \Lambda_\gamma^*(\xi) < \infty, \\ \exp\Big(-\dfrac{1}{\epsilon\delta}\Big), & \text{if } \Lambda_\gamma^*(\xi) = \infty. \end{cases}$$

To prove the relation (12), let $\ell = \inf_K \Lambda_\gamma^*$. Since K is a compact of $S'(\mathbb{R})$ (and by definition of a compact set in dual space of nuclear space see [5]), we can choose $\xi_1, \xi_2, \ldots, \xi_n \in K$, $r_1, r_2, \ldots, r_n \in \mathbb{R}_+^*$ and $p_1, p_2, \ldots, p_n \in \mathbb{N}^*$, so that $K \subset \bigcup_{k=1}^n B_{p_k}(\xi_k, r_k)$, and for $k \in \{1, 2, \ldots, n\}$, $B_{p_k}(\xi_k, r_k)$ satisfies the inequality (11), then

$$\gamma_\epsilon(K) \leq n \exp\Big(-\frac{1}{\epsilon} \min\Big(\frac{1}{\delta}, \ell - \delta\Big)\Big)$$

and therefore

$$\limsup_{\epsilon \to 0} \epsilon \log(\gamma_\epsilon(K)) \leq -\min\Big(\frac{1}{\delta}, \ell - \delta\Big).$$

Finally, let $\delta \searrow 0$, then we obtain

$$\limsup_{\epsilon \to 0} \epsilon \log(\gamma_\epsilon(K)) \leq -\inf_K \Lambda_\gamma^*.$$

□

Together with lemma 3.2 we need the following lemma to prove the inequality (1).

Lemma 3.3. *There exists some $m > 0$, $p \in \mathbb{N}^*$ so that*

$$\int_{S_{-p}(\mathbb{R})} \exp\Big(\frac{m|y|_{-p}^2}{2}\Big) d\gamma(y) < \infty. \tag{13}$$

Moreover, the set $K_L = \{y \in S'(\mathbb{R}); \frac{1}{2}|y|_{-p}^2 \leq \frac{L}{m}\}$ is a compact subset of $S'(\mathbb{R})$ for each $L > 0$ and

$$\limsup_{\epsilon \to 0} \epsilon \log(\gamma_\epsilon(K_L^C)) \leq -L, \tag{14}$$

where K_L^C is the complement of K in $S'(\mathbb{R})$.

Proof. The relation (13) is an immediate consequence of the integrability condition (4) for the particular case where μ is the Gaussian measure γ. For all $\epsilon > 0$

$$\begin{aligned}\gamma_\epsilon(K_L^C) &= \gamma_\epsilon\Big(\Big\{y \in S'(\mathbb{R}), \frac{m}{2}|y|_{-p}^2 > L\Big\}\Big)\\ &= \gamma\Big(\Big\{y \in S'(\mathbb{R}), \frac{m}{2}|y|_{-p}^2 > \frac{L}{\epsilon}\Big\}\Big)\\ &= \int_{\{y\in S'(\mathbb{R}), \frac{m}{2}|y|_{-p}^2 > \frac{L}{\epsilon}\}} e^{-\frac{m}{2}|x|_{-p}^2}\, e^{\frac{m}{2}|x|_{-p}^2}\, d\gamma(x)\\ &\leq \exp\Big(-\frac{L}{\epsilon}\Big)\int_{S_{-p}} e^{\frac{m}{2}|x|_{-p}^2}\, d\gamma(x)\,.\end{aligned}$$

Together with (13), this surely leads to (14). □

Proposition 3.1. *Let F a closed set of $S'(\mathbb{R})$; we have*

$$\limsup_{\epsilon\longrightarrow 0} \epsilon \log(\gamma_\epsilon(F)) \leq -\inf_F(\Lambda_\gamma^*). \tag{15}$$

Proof. Let $\ell = \inf_F \Lambda_\gamma^*$, and for $L > 0$ set $F_L = F \bigcap K_L$, where K_L is the compact set produced in the lemma 3.3. Then:

$$\gamma_\epsilon(F) \leq \gamma_\epsilon(F_L) + \gamma_\epsilon(K_L^C)$$

and so by lemma 3.2 and lemma 3.3, we have

$$\overline{\lim_{\epsilon\longrightarrow 0}}\, \epsilon \log(\gamma_\epsilon(F)) \leq -\min(\ell, L).$$

After letting $L \longrightarrow \infty$, we obtain the desire results. □

To prove the inequality (2) for a sequence of Gaussian measures, we need the following result concerning the quasi invariance of the Gaussian measure γ on $S'(\mathbb{R})$; see for example [6], [9] and [12].

Lemma 3.4. *Given $\xi \in L^2(\mathbb{R})$, let γ^ξ denote the image measure of γ under the map $x \longmapsto x + \xi$. So*

$$\frac{d\gamma^\xi(x)}{d\gamma(x)} := \frac{\gamma(dx-\xi)}{\gamma(dx)} = e^{\langle x,\xi\rangle - \frac{1}{2}\langle \xi,\xi\rangle}, \quad x \in S'(\mathbb{R}),$$

i.e., the Gaussian measure γ is quasi-invariant under the translation by any $\xi \in L^2(\mathbb{R})$ and $\frac{\gamma(dx-\xi)}{\gamma(dx)}$ in the Radon-Nikodym derivative.

Proposition 3.2. *For every open $G \subset S'(\mathbb{R})$,*

$$\liminf_{\epsilon\to 0} \epsilon \log(\gamma_\epsilon(G)) \geq -\inf_G \Lambda_\gamma^* . \tag{16}$$

Proof. For every open $G \subset S'(\mathbb{R})$, there exist $r > 0$, $p > 0$ and $\xi \in S_{-p}(\mathbb{R})$ so that $B_p(\xi, r) \subseteq G$. Since the space $L^2(\mathbb{R})$ is densely imbedded in $S'(\mathbb{R})$ for the weak topology, there exist $(\xi_n)_n \in L^2(\mathbb{R})$ so that $\xi = \lim_{n\to\infty} \xi_n$ weakly. By the lower semi-continuity property of the function Λ_γ^*, we have:

$$\lim_{n\to\infty} \Lambda_\gamma^*(\xi_n) = \Lambda_\gamma^*(\xi).$$

Hence, we need only to prove (16) for $\xi \in L^2(\mathbb{R})$.

$$\begin{aligned}
\gamma_\epsilon(G) \geq \gamma_\epsilon(B_p(\xi, r)) &= \gamma\Big(B_p\Big(\frac{\xi}{\sqrt{\epsilon}}, \frac{r}{\sqrt{\epsilon}}\Big)\Big) \\
&= \int_{B_p\left(0, \frac{r}{\sqrt{\epsilon}}\right)} d\gamma\Big(x - \frac{\xi}{\sqrt{\epsilon}}\Big) \\
&= \int_{B_p\left(0, \frac{r}{\sqrt{\epsilon}}\right)} e^{\left\langle x, \frac{\xi}{\sqrt{\epsilon}}\right\rangle - \frac{1}{2\epsilon}(\xi,\xi)}\, d\gamma(x) \\
&\geq \exp\Big(-\frac{1}{\epsilon}\big(r|\xi|_{-p} + \Lambda_\gamma(\xi)\big)\Big) \int_{B_p\left(0, \frac{r}{\sqrt{\epsilon}}\right)} d\gamma(x).
\end{aligned}$$

Since, for $\xi \in L^2(\mathbb{R})$, $\Lambda_\gamma(\xi) = \Lambda_\gamma^*(\xi)$, we see that the relation (16) holds. □

Propositions 3.1 and 3.2 prove in the following theorem that the family $\{\gamma_\epsilon, \epsilon > 0\}$ satisfies the full large deviation principle with rate function Λ_γ^*.

Theorem 3.1. *For every measurable Γ in $S'(\mathbb{R})$, we have*

$$-\inf_{\Gamma^0}(\Lambda_\gamma^*) \leq \liminf_{\epsilon\to 0} \epsilon \log(\gamma_\epsilon(\Gamma)) \leq \limsup_{\epsilon\to 0} \epsilon \log(\gamma_\epsilon(\Gamma)) \leq -\inf_{\overline{\Gamma}}(\Lambda_\gamma^*) . \tag{17}$$

In particular, for a given $\xi \in S(\mathbb{R})$ and $x \in \mathbb{R}$, consider the half-plan in $S'(\mathbb{R})$ associated to ξ and x, given by (5). Then we obtain the following result:

Corollary 3.1. *For a given $\xi \in S(\mathbb{R})$ and $x > 0$, there exist $p > 0$ such that*

$$\begin{aligned}
-\inf_{y\in A_{\xi,x}} \sup_{\lambda\in S(\mathbb{R})} \frac{1}{2}\Big(\frac{\langle y, \lambda\rangle}{|\lambda|_0}\Big)^2 &\leq \liminf_{\epsilon\to 0} \epsilon \log(\gamma_\epsilon(A_{\xi,x})) \\
&\leq \limsup_{\epsilon\to 0} \epsilon \log(\gamma_\epsilon(A_{\xi,x})) \leq -\frac{1}{2}\frac{x^2}{|\xi|_p^2} ,
\end{aligned} \tag{18}$$

where $A_{\xi,x} = \{y \in S'(\mathbb{R}); \langle y, \xi\rangle > x\}$.

Proof. Using the definition of Λ^*_γ, the relation (9) and the fact that the topology on $S(\mathbb{R})$ is defined by a family $\{|.|_p, p \in \mathbb{N}\}$ of increasing Hilbertian norms, we note that for all $q \in \mathbb{N}$

$$\Lambda^*_\gamma(y) \geq \frac{1}{2}|y|^2_{-q}, \quad \text{for all } y \in S'(\mathbb{R}), \tag{19}$$

and there exists $p \in \mathbb{N}$ such that

$$|\langle y, \xi \rangle| \leq |y|_{-p}\,|\xi|_p, \quad \text{for all } y \in S'(\mathbb{R}). \tag{20}$$

Combining the equations (19), (20) and the definition of $A_{\xi,x}$ we prove the right inequality of the equation (18). To prove the left inequality of equation (18), we observe that for all $\lambda \in S(\mathbb{R})$, $y \in S'(\mathbb{R})$:

$$\langle y, \lambda \rangle - \frac{1}{2}|\lambda|_0^2 \leq \frac{1}{2}\,\frac{\langle y, \lambda \rangle^2}{|\lambda|_0^2},$$

so we obtain

$$\Lambda^*_\gamma(y) \leq \sup_{\lambda \in S(\mathbb{R})} \frac{1}{2}\left(\frac{\langle y, \lambda \rangle}{|\lambda|_0}\right)^2;$$

finally we obtain the desired results. □

Remark 3.1.

1. Theorem 2.2 obtained in [15] gives for the Gaussian measure γ_ϵ the following tail estimate:

$$\exists p \in \mathbb{N} : \gamma_\epsilon(A_{\xi,x}) \leq C \exp\left(-\frac{1}{2}\,\frac{x^2}{\epsilon|\xi|_p^2}\right), \tag{21}$$

which implies that

$$\liminf_{\epsilon \to 0} \epsilon \log(\gamma_\epsilon(A_{\xi,x})) \leq \limsup_{\epsilon \to 0} \epsilon \log(\gamma_\epsilon(A_{\xi,x})) \leq -\frac{1}{2}\,\frac{x^2}{|\xi|_p^2}. \tag{22}$$

So the inequality (22) is only the right hand inequality of (18). Therefore corollary 3.1 generalizes and precises the result obtained in [15]. In fact, the image measure of γ_a by the map (10) is given by $\gamma_{a\epsilon} = \gamma_{\epsilon'}$. So if $\epsilon \to 0$ then $\epsilon' \to 0$ and

$$\Lambda^*_{\gamma_a}(\varphi) = \begin{cases} \frac{1}{2a}|\varphi|_0^2, & \text{if } \varphi \in L^2(\mathbb{R}, dx), \\ +\infty, & \text{if } \varphi \in S'(\mathbb{R}) \backslash L^2(\mathbb{R}, dx). \end{cases}$$

2. Analogously we recover the same results if we replace the measure γ by the Gaussian measure γ_a defined on $S'(\mathbb{R})$ with mean 0 and variance $a > 0$, i.e.,

$$\int_{S'(\mathbb{R})} e^{i\langle y,\xi\rangle}\, d\gamma_a(y) = e^{-\frac{1}{2}a|\xi|_0^2} = e^{-\frac{1}{2}a(\xi,\xi)}, \quad \xi \in S(\mathbb{R}), \quad a > 0.$$

3. In the particular case where γ_ϵ is the Gaussian measure with mean 0 and variance ϵ on $\mathbb{R}$, the large deviation principle given in (18) becomes the following equality:

$$\lim_{\epsilon\to 0} \epsilon \log(\gamma_\epsilon(A_{\xi,x})) = -\frac{x^2}{2\xi^2}.$$

4. Application to stochastic differential equations

4.1. Generalized Gross heat equation

It is well known that in infinite dimensional complex analysis the convolution operator on a general function space is defined as a continuous operator which commutes with the translation operator.

Let us define the convolution $\Phi * \varphi$ of a distribution $\Phi \in \mathcal{F}_\theta(N')^*$ and a test function $\varphi \in \mathcal{F}_\theta(N')$ to be the function

$$(\Phi * \varphi)(x) = \langle\langle \Phi, t_{-x}\varphi\rangle\rangle, \quad x \in N', \tag{23}$$

where $t_x\varphi$ is the translation operator, i.e.,

$$t_x\varphi(y) = \varphi(y - x), \quad y \in N'.$$

Note that $\Phi * \varphi \in \mathcal{F}_\theta(N')$ for any $\varphi \in \mathcal{F}_\theta(N')$ and the convolution product is given in terms of the dual pairing as $(\Phi * \varphi)(0) = \langle\langle \Phi, \varphi\rangle\rangle$ for any $\Phi \in \mathcal{F}_\theta(N')^*$ and $\varphi \in \mathcal{F}_\theta(N')$. We can generalize the above convolution product for generalized functions as follows. Let $\Phi, \Psi \in \mathcal{F}_\theta(N')^*$ be given, then $\Phi * \Psi$ is defined by

$$\langle\langle \Phi * \Psi, \varphi\rangle\rangle = \langle\langle \Phi, \Psi * \varphi\rangle\rangle, \quad \forall \varphi \in \mathcal{F}_\theta(N'). \tag{24}$$

The Gross Laplacian $\Delta_G\varphi(x)$ of φ at $x \in N'$ is given by

$$\Delta_G\varphi(x) = \sum_{n\geq 0} (n+2)(n+1)\langle x^{\otimes n}, \langle \tau, \varphi^{(n+2)}\rangle\rangle,$$

for $\varphi \in \mathcal{F}_\theta(N')$ represented by $\varphi(x) = \sum_{n\geq 0}\langle x^{\otimes n}, \varphi^{(n)}\rangle$ and τ is the trace operator defined by $\langle \tau, \xi \otimes \eta\rangle = \langle \xi, \eta\rangle$, $\xi, \eta \in N$. For more information on the Gross Laplacian, see [7],[8],[10] and [12]. In fact, the Gross Laplacian Δ_G is a convolution operator given by

$$\Delta_G(\Psi) = \mathcal{T} * \Psi \tag{25}$$

where $\mathcal{T}$ is the distribution in $\mathcal{F}_\theta(N')^*$ such that its Laplace transform is given by $\widehat{\mathcal{T}}(z) = \langle \tau, z^{\otimes 2}\rangle$.

Theorem 4.1. *[1] Let θ be a Young function satisfying $\lim_{r\to+\infty}\theta(r)/r^2 < \infty$ and $F \in \mathcal{F}_\theta(N')^*$. Then the following generalized Gross heat equation perturbed by the white noise $\dot{W}_t$*

$$\frac{\partial U_t}{\partial t} = \frac{1}{2}\Delta_G U_t + \alpha \dot{W}_t, \quad t \geq 0, \quad U(0) = F,\ \alpha \in \mathbb{R}, \tag{26}$$

has a unique solution in $\mathcal{F}_\theta(N')^*$ given by

$$U_t = F * \widetilde{\gamma}_t + \alpha \int_0^t \widetilde{\gamma}_{t-s} * \dot{W}_s\, ds\,;$$

$\widetilde{\gamma}_t$ is a positive distribution in $\mathcal{F}_\theta(N')^*$ given by

$$\langle\langle \widetilde{\gamma}_t, \varphi\rangle\rangle = \int_{X'} \varphi(x)\, d\gamma_t(x), \quad \textit{for all } \varphi \in \mathcal{F}_\theta(N'),$$

where $\widehat{\gamma}_t(\xi) = e^{\frac{1}{2}t\langle \xi,\xi\rangle}$, $\forall \xi \in X$.

From theorems 4.1 and 3.1 we obtain the following result:

Corollary 4.1. *Let μ_{U_t} be the associated measure with the solution of Cauchy problem (26) for the particular case where $\alpha = 0$, the Young function θ is given by $\theta(x) = x^2/2$ and F is a the standard Gaussian distribution on $S'(\mathbb{R})$. Then the family of measures $\{\mu^\epsilon_{U_t}, \epsilon > 0\}$ of image measure of the measure μ_{U_t} under the map (10) satisfies the full large deviation principle (17) with rate function $\Lambda^*_{\mu_{U_t}}$.*

Proof. If $\alpha = 0$, the Young function θ is given by $\theta(x) = x^2/2$ and F is the standard Gaussian distribution on $S'(\mathbb{R})$, then the solution of (26) is a positive generalized function and given by the explicit formula

$$U_t = \widetilde{\gamma}_{t+1} \in \mathcal{F}_\theta(S'(\mathbb{R}))^*_+.$$

So theorem 2.1 guarantees the existence and uniqueness of a Radon measure μ_{U_t} on $S'(\mathbb{R})$ associated with U_t such that

$$\widehat{\mu}_{U_t}(\xi) = e^{\frac{1}{2}(1+t)|\xi|_0^2}.$$

Therefore, the desired result is a consequence from theorem 3.1. □

4.2. Langevin Equation

We can apply Schilder's theorem for the measure associated with the solution of Langevin equation. In fact, the following Langevin equation

$$\begin{cases} dV_t = -aV_t dt + \sigma dW_t, & a > 0,\ \sigma > 0, \\ V(0) = V_0, \end{cases} \tag{27}$$

where W_t is the Wiener process, has a unique solution given by:

$$V_t = V_0 e^{-at} + \sigma \int_0^t e^{-a(t-s)}\, dW_s\,.$$

The process V_t is called the Ornstein-Uhlenbeck process, and if V_0 has a Gaussian distribution and is independent of W, V_t is a Gaussian distribution with parameters:

$$\begin{aligned} \mathrm{E}(V_t) &= \mathrm{E}(V_0)\, e^{-at}\,, \\ \mathrm{Var}(V_t) &= \mathrm{Var}(V_0)\, e^{-2at} + \frac{\sigma^2}{2a}(1 - e^{-2at})\,. \end{aligned}$$

In particular, if V_0 has a Gaussian distribution independent of W with mean 0 and variance $\frac{\sigma^2}{2a}$, the solution of (27) has a Gaussian distribution with mean 0 and variance $\frac{\sigma^2}{a}$. And the family $\{\mu_{V_t}^{\epsilon}\}$ of image measures of μ_{V_t} under the map (10) satisfies the full large deviation principle (17).

4.3. Ventcel and Freidlin's estimate

Our third application of Schilder's theorem will be to Ventcel and Freidlin's estimate on the large deviations of randomly perturbed dynamical systems (see [18]). The theory of Ventcel and Freidlin deals with families of measures $\{P_\epsilon : \epsilon > 0\}$ on $S'(\mathbb{R})$ of which the following is a typical example. For a given bounded, uniformly Lipschitz continuous function $b : \mathbb{R} \to \mathbb{R}$, define the map $X : S'(\mathbb{R}) \mapsto S(\mathbb{R})$ by

$$X(t,\xi) = \xi(t) + \int_0^t b(X(s,\xi))\, ds, \quad t \in \mathbb{R},\ \xi \in S(\mathbb{R}).$$

This integration equation is equivalent to the following stochastic differential equation:

$$\begin{cases} dX_t(\xi) = d\xi(t) + f(X_t(\xi))\, ds\,, \\ X(0,\xi) = \xi(0)\,. \end{cases}$$

For $\epsilon > 0$, let $P_\epsilon = \gamma_\epsilon \circ X^{-1}$ be the image measure of γ_ϵ under the map $\xi \mapsto X(\xi)$. It is easy to see that the map $\xi \in S^{'}(\mathbb{R}) \mapsto X(\xi) \in S(\mathbb{R})$ is a continuous one, see [13]. Thus, if $G \subseteq S(\mathbb{R})$ is open, then Schilder's theorem says that

$$\begin{aligned} \liminf_{\epsilon\to 0} \epsilon \ln(P_\epsilon(G)) &= \liminf_{\epsilon\to 0} \epsilon \ln\Big(\gamma_\epsilon(X^{-1}(G))\Big) \\ &= \liminf_{\epsilon\to 0} \epsilon \ln\Big(\gamma_\epsilon(\{\xi \in S'(\mathbb{R}) : X(\xi) \in G\})\Big) \\ &\geq -\inf\{\Lambda^*_\gamma(\xi),\ X(\xi) \in G\} \\ &= -\inf_G(\Lambda^*_\gamma \circ X^{-1}) = -\inf_{G'}(\Lambda^*_\gamma), \end{aligned}$$

where $G' = X^{-1}(G) \subset S'(\mathbb{R})$ which is also an open set. Similarly, if $F \subseteq S(\mathbb{R})$ is closed, then

$$\limsup_{\epsilon\to 0} \epsilon \log(\gamma_\epsilon(F)) \leq -\inf_{F'}(\Lambda^*_\gamma),$$

where $F' = X^{-1}(F) \subset S'(\mathbb{R})$.

References

1. A. Barhoumi, H. H. Kuo, H. Ouerdiane, "Generalized Gross Heat equation with noises", *Soochow Journal of Mathematics* **32**, no. 1, 113–125 (2006).
2. M. Ben Chrouda, M. El Oued, H. Ouerdiane, "Convolution calculus and applications to stochastic differential equations", *Soochow Journal of Mathematics* **28**, no. 4, 345–388 (2002).
3. H. Cramer, "Sur un nouveau théorème-limite de la théorie des probabilités", *Actualités Scientifiques et Industrielles*, 736, 5–23 (1938), *Colloque consacré à la théorie des probabilités*, vol. 3, Hermann, Paris.
4. R. Gannoun, R. Hachaichi, H. Ouerdiane, A. Rezgui, "Un théorème de dualité entre espaces de fonctions holomorphes à croissance exponentielle", *J. Func. Anal.* **171**(1), 1–14 (2000).
5. I. M. Gelfand, N. Ya. Vilenkin, *Generalized Functions*, vol. IV, Academic Press, New York and London, 1968.
6. N. Obata, *White Noise Calculus and Fock Space*, vol. 1577, L.N.M Springer-Verlag, Berlin, Heidelberg, and New York, 1994.
7. L. Gross, P. Maliavin, "Hall's transformation and the Segal-Bargmann map", in *Itô's Stochastic Calculus and Probability*, Springer-Verlag, Tokyo, 1996, pp. 73–116.

8. T. Hida, *Brownian Motion*, Springer-Verlag, New York, 1980.
9. T. Hida, H. H. Kuo, J. Potthoff, L. Streit, *White Noise And Infinite Dimensional Calculus*, Kluwer, Dordrecht, 1993.
10. H. Holden, B. Øksendal, J. Ubøe, T. Zhang, *Stochastic Partial Differential Equations: A Modeling White Noise Approach* Birkhäuser, Boston, Basel, Berlin, 1996.
11. Jean-Dominique Deuschel, Daniel W. Stroock, *Large Deviations*, AMS Chelsea Publishing, American Mathematical Society, . Providence, Rhode Island, 2001.
12. H. H. Kuo, *White Noise Distribution Theory*, CRC Press, Boca Raton, 1996.
13. H. H. Kuo, *Introduction to Stochastic Integration*, Universitex, Springer, 2006.
14. H. Ouerdiane, A. Rezgui, "Un théorème de Bochner-Minlos avec une condition d'intégrabilité", *Infinite Dimensional Analysis, Quantum Probability and Related Topics* **3**, no. 2, 297–302 (2000).
15. H. Ouerdiane, N. Prilvaut, "Asymptotics estimates for white noise distributions", *C. R. Acad. Sci. Paris, Ser. I* **338**, 799–804 (2004).
16. M. Schilder, "Some asymptotics formulae for Wiener integrals", *Trans. Amer. Math. Soc.* **125**, 63–85 (1966).
17. D. W. Stroock, *An Introduction to the Theory of Large Deviation*, Universitex, Springer-Verlag, 1984.
18. S. R. S. Varadhan, "Asymptotic probabilities and differential equations", *Comm. Pure Appl. Math.* **19**, 261–286 (1966).

Sonia Chaari
University of Tunis El Manar, Faculty of Sciences of Tunis, Department of Mathematics, Campus Universitaire, 1060 Tunis, Tunisia
Sonia.Chaari@fsb.rnu.tn

Soumaya Gheryani
University of Tunis El Manar, Faculty of Sciences of Tunis, Department of Mathematics, Campus Universitaire, 1060 Tunis, Tunisia
soumaye_gheryani@yahoo.fr

Habib Ouerdiane
University of Tunis El Manar, Faculty of Sciences of Tunis, Department of Mathematics, Campus Universitaire, 1060 Tunis, Tunisia
habib.ouerdiane@fst.rnu.tn

A NONLINEAR STOCHASTIC EQUATION OF CONVOLUTION TYPE

F. Cipriano (*GFM / FCT-UNL, Lisboa*),
H. Ouerdiane (*Faculté des Sciences de Tunis*),
J. L. Silva (*University of Madeira*),
R. Vilela Mendes (*CMAF, Lisboa*)

A nonlinear equation, similar to the Burgers equation with the usual product replaced by a convolution product, is studied. The initial condition is a generalized function. By using the Laplace transform in a general white noise analysis setting, a general solution is found in $(1+n)$ dimensions.

1. Introduction

The aim of this paper is to study the following nonlinear stochastic equation of convolution type

$$\begin{cases} \dfrac{\partial \vec{u}}{\partial t} + (\vec{u} * \vec{\nabla})\vec{u} = \nu \Delta \vec{u} + \vec{f} * \vec{u}, \\ \vec{u}(0, x) = \vec{u}_0(x), \end{cases} \tag{1}$$

where $\vec{u}$ is a $\mathbb{R}^n$-generalized vector field, $\vec{f}$ is a n-dimensional generalized function, $\nu > 0$ a real constant, $t \in [0, \infty)$ the time parameter, $x = (x_1, \ldots, x_n) \in \mathbb{R}^n$ the spatial variable, Δ the Laplacian operator in $\mathbb{R}^n$, ∇ the gradient and $*$ the convolution product for generalized functions (see [1], [11] and subsection 2.2 for more details), $\vec{f} * \vec{u}$ is a driving term, $\vec{u} * \vec{\nabla}$ denotes the differential operator

$$\sum_{j=1}^{n} u_j * \frac{\partial}{\partial x_j}$$

and the initial condition $\vec{u}_0 = (u_{0,1}, \ldots, u_{0,n})$ is a n-dimensional generalized function, see section 3 for more details.

Problem (1) with $*$ replaced by the usual product coincides with the classical Burgers equation well known in the literature ([3], [5] and references therein). However the physical interpretation of (1) is quite different. This is easily seen by comparing the j-component, for any j, $1 \leq j \leq n$, the Fourier transform of the nonlinear kinetic term $(\vec{u}.\nabla)\vec{u}$ in the Burgers equation, namely

$$\int_{\mathbb{R}^n} d^n q \sum_{i=1}^{n} \tilde{u}_i(t,q)(k-q)_i \tilde{u}_j(t,k-q) \tag{2}$$

with the j-component of the Fourier transform of the nonlinear term in (1), $(\vec{u} * \nabla)\vec{u}$

$$\sum_{i=1}^{n} \tilde{u}_i(t,k) k_i \tilde{u}_j(t,k), \tag{3}$$

$\tilde{u}_i$ denoting the Fourier transform of u_i.

In the Burgers case the expression (2) implies that the Fourier modes at length scale $\frac{2\pi}{q}$ and $\frac{2\pi}{k-q}$ control the eddies at scale $\frac{2\pi}{k}$, consistent with the phenomenological description of the inertial range in the turbulence cascade. However, in the convolution case the nonlocal nonlinearity corresponds to a self-interaction of the modes at each length scale. Nevertheless, nonlocal nonlinearities are also important in models of transport in magnetized plasmas, see [4], and also in the modeling of convection driven by density gradients as it arises in geophysical fluid flows, see [12], [13], [14] and [15].

By restricting oneself to solutions of gradient type, the Burgers equation may be linearized by the Cole-Hopf transformation. This provides the most general solution for $(1+1)$ dimensions but not for $(1+n)$ dimensions. For our equation (1), using the Laplace transform in a general setting, we obtain a general solution for $(1+n)$ dimensions.

The paper is organized as follows: In Section 2 we provide the mathematical background needed to solve the Cauchy problem stated above, namely spaces of test and generalized functions, the characterization theorem of generalized functions and the convolution product as well as some of its properties. In Section 3 we combine the convolution calculus and the characterization theorem in order to find an explicit solution of the problem (1).

2. Preliminaries

2.1. Test and generalized functions spaces

In this section we introduce the framework needed later on. The starting point is the real Hilbert space $\mathcal{H} = L^2(\mathbb{R}, \mathbb{R}^d) \times \mathbb{R}^r$, $d, r \in \mathbb{N}$ with scalar product $(\cdot,\cdot)$ and norm $|\cdot|$. More precisely, if

$$(f,x) = ((f_1,\ldots,f_d),(x_1,\ldots,x_r)) \in \mathcal{H},$$

then

$$|(f,x)|^2 := \sum_{i=1}^{d} \int_{\mathbb{R}} f_i^2(u)du + \sum_{i=1}^{r} x_i^2 = |f|_{L^2}^2 + |x|_{\mathbb{R}^r}^2.$$

Let us consider the real nuclear triplet

$$\mathcal{M}' = S'(\mathbb{R},\mathbb{R}^d) \times \mathbb{R}^r \supset \mathcal{H} \supset S(\mathbb{R},\mathbb{R}^d) \times \mathbb{R}^r = \mathcal{M}. \tag{4}$$

The pairing $\langle\cdot,\cdot\rangle$ between $\mathcal{M}'$ and $\mathcal{M}$ is given in terms of the scalar product in $\mathcal{H}$, i.e., $\langle(\omega,x),(\xi,p)\rangle := (\omega,\xi)_{L^2} + (x,p)_{\mathbb{R}^r}$, $(\omega,x) \in \mathcal{M}'$ and $(\xi,p) \in \mathcal{M}$. Since $\mathcal{M}$ is a Fréchet nuclear space, it can be represented as

$$\mathcal{M} = \bigcap_{n=0}^{\infty} S_n(\mathbb{R},\mathbb{R}^d) \times \mathbb{R}^r = \bigcap_{n=0}^{\infty} \mathcal{M}_n,$$

where $S_n(\mathbb{R},\mathbb{R}^d) \times \mathbb{R}^r$ is a Hilbert space with norm squared given by $|\cdot|_n^2 + |\cdot|_{\mathbb{R}^r}^2$, see e.g. [8] or [2] and references therein. We will consider the complexification of the triple (4) and denote it by

$$\mathcal{N}' \supset \mathcal{Z} \supset \mathcal{N}, \tag{5}$$

where $\mathcal{N} = \mathcal{M} + i\mathcal{M}$ and $\mathcal{Z} = \mathcal{H} + i\mathcal{H}$. On $\mathcal{M}'$ we have the standard Gaussian measure γ given by Minlos's theorem via its characteristic functional, namely for every $(\xi,p) \in \mathcal{M}$

$$C_\gamma(\xi,p) = \int_{\mathcal{M}'} \exp\big(i\langle(\omega,x),(\xi,p)\rangle\big)\, d\gamma((\omega,x)) = \exp\Big(-\frac{1}{2}(|\xi|_{L^2}^2 + |p|_{\mathbb{R}^r}^2)\Big).$$

In order to solve the $(1+n)$-dimensional equation of convolution type (1) we need to introduce an appropriate space of vectorial generalized functions. We borrow this construction from [9]. Let $\theta = (\theta_1,\theta_2) : \mathbb{R}_+^2 \to \mathbb{R}$, $(t_1,t_2) \mapsto \theta_1(t_1) + \theta_2(t_2)$ where θ_1, θ_2 are two Young functions, i.e., $\theta_i : \mathbb{R}_+ \to \mathbb{R}_+$ continuous convex strictly increasing function and

$$\lim_{t\to\infty} \frac{\theta_i(t)}{t} = \infty, \quad \theta_i(0) = 0, \quad i = 1,2.$$

For every pair $m = (m_1,m_2)$ with $m_1, m_2 \in]0,\infty[$, we define the Banach space $\mathcal{F}_{\theta,m}(\mathcal{N}_{-n})$, $n \in \mathbb{N}$, by

$$\mathcal{F}_{\theta,m}(\mathcal{N}_{-n}) := \big\{f : \mathcal{N}_{-n} \to \mathbb{C}, \text{ entire}, \|f\|_{\theta,m,n} < \infty\big\},$$

where

$$\|f\|_{\theta,m,n} := \sup_{z\in\mathcal{N}_{-n}} \{|f(z)| \exp(-\theta(m|z|_{-n}).$$

Here, for each $z = (\omega, x)$ we have $\theta(m|z|_{-n}) := \theta_1(m_1|\omega|_{-n}) + \theta_2(m_2|x|)$. Now we consider as test function space the space of entire functions on $\mathcal{N}'$ of (θ_1, θ_2)-exponential growth and minimal type

$$\mathcal{F}_\theta(\mathcal{N}') := \bigcap_{m\in(\mathbb{R}_+^*)^2, n\in\mathbb{N}_0} \mathcal{F}_{\theta,m}(\mathcal{N}_{-n}),$$

endowed with the projective limit topology. We would like to use $\mathcal{F}_\theta(\mathcal{N}')$ to construct a triple centered in the complex Hilbert space $L^2(\mathcal{M}', \gamma)$. To this end we need another condition on the pair of Young functions (θ_1, θ_2). Namely,

$$\lim_{t\to\infty} \frac{\theta_i(t)}{t^2} < \infty, \quad i = 1, 2. \tag{6}$$

This is enough to obtain the following Gelfand triple

$$\mathcal{F}_\theta'(\mathcal{N}') \supset L^2(\mathcal{M}', \gamma) \supset \mathcal{F}_\theta(\mathcal{N}'), \tag{7}$$

where $\mathcal{F}_\theta'(\mathcal{N}')$ is the topological dual of $\mathcal{F}_\theta(\mathcal{N}')$ with respect to $L^2(\mathcal{M}', \gamma)$ endowed with the inductive limit topology.

In applications it is very important to have the characterization of generalized functions from $\mathcal{F}_\theta'(\mathcal{N}')$. First we define the Laplace transform of an element in $\mathcal{F}_\theta'(\mathcal{N}')$. For every fixed element $(\xi, p) \in \mathcal{N}$ the exponential function $\exp((\xi, p))$ is a well defined element in $\mathcal{F}_\theta(\mathcal{N}')$, see [7]. The Laplace transform $\mathcal{L}$ of a generalized function $\Phi \in \mathcal{F}_\theta'(\mathcal{N}')$ is defined by

$$\hat{\Phi}(\xi, p) := (\mathcal{L}\Phi)(\xi, p) := \langle\!\langle \Phi, \exp((\xi, p))\rangle\!\rangle. \tag{8}$$

We are ready to state the characterization theorem (see, e.g., [7] and [1] for the proof) which is the main tool in our further consideration.

Theorem 2.1.

1. *The Laplace transform is a topological isomorphism between $\mathcal{F}_\theta'(\mathcal{N}')$ and the space $\mathcal{G}_{\theta^*}(\mathcal{N})$, where $\mathcal{G}_{\theta^*}(\mathcal{N})$ is defined by*

$$\mathcal{G}_{\theta^*}(\mathcal{N}) := \bigcup_{m\in(\mathbb{R}_+^*)^2, n\in\mathbb{N}_0} \mathcal{G}_{\theta^*,m}(\mathcal{N}_n),$$

and $\mathcal{G}_{\theta^*,m}(\mathcal{N}_n)$ is the space of entire functions on $\mathcal{N}_n$ with the following θ-exponential growth condition

$$\mathcal{G}_{\theta^*,m}(\mathcal{N}_n) \ni g, \quad |g(\xi,p)| \leq k \exp(\theta_1^*(m_1|\xi|_n) + \theta_2^*(m_2|p|)), \qquad (\xi,p) \in \mathcal{N}_n.$$

Here $\theta^* = (\theta_1^*, \theta_2^*)$, where $\theta_i^*(x) = \sup_{t\geq 0}(tx - \theta_i(t))$ is the Legendre transform associated to the function θ_i, $i = 1, 2$.

2. *In the particular case $\theta(x) = (\theta_1(x), \theta_2(x)) = (x, x)$, we denote the space $\mathcal{F}'_\theta(\mathcal{N}')$ by $\mathcal{F}'_x(\mathcal{N}')$. Then the Laplace transform realizes a topological isomorphism between the distributions space $\mathcal{F}'_x(\mathcal{N}')$ and the space* $\mathrm{Hol}_0(\mathcal{N})$ *of holomorphic functions on a neighborhood of zero of $\mathcal{N}$.*

2.2. The Convolution Product $*$

It is well known that in infinite dimensional complex analysis the convolution operator on a general function space $\mathcal{F}$ is defined as a continuous operator which commutes with the translation operator. Let us define the convolution between a generalized and a test function. Let $\Phi \in \mathcal{F}'_\theta(\mathcal{N}')$ and $\varphi \in \mathcal{F}_\theta(\mathcal{N}')$ be given, then the convolution $\Phi * \varphi$ is defined by

$$(\Phi * \varphi)(\omega, x) := \langle\langle \Phi, \tau_{-(\omega,x)}\varphi \rangle\rangle,$$

where $\tau_{-(\omega,x)}$ is the translation operator, i.e.,

$$(\tau_{-(\omega,x)}\varphi)(\eta, y) := \varphi(\omega + \eta, x + y).$$

It is not hard the see that $\Phi * \varphi \in \mathcal{F}_\theta(\mathcal{N}')$, cf. [7]. The convolution product is given in terms of the dual pairing as $(\Phi * \varphi)(0,0) = \langle\langle \Phi, \varphi \rangle\rangle$ for any $\Phi \in \mathcal{F}'_\theta(\mathcal{N}')$ and $\varphi \in \mathcal{F}_\theta(\mathcal{N}')$.

We can generalize the above convolution product for generalized functions as follows. Let $\Phi, \Psi \in \mathcal{F}'_\theta(\mathcal{N}')$ be given, then $\Phi * \Psi$ is defined by

$$\langle\langle \Phi * \Psi, \varphi \rangle\rangle := \langle\langle \Phi, \Psi * \varphi \rangle\rangle, \ \forall \varphi \in \mathcal{F}_\theta(\mathcal{N}'). \tag{9}$$

This definition of convolution product for generalized functions will be used later for the solution of the equation (1). We have the following equality, (see [11, Proposition 3]):

$$\Phi * \exp((\xi, p)) = (\mathcal{L}\Phi)(\xi, p) \exp((\xi, p)), \quad (\xi, p) \in \mathcal{N}.$$

As a consequence of the above equality and definition (9) we obtain

$$\mathcal{L}(\Phi * \Psi) = \mathcal{L}\Phi\mathcal{L}\Psi, \;\; \Phi, \Psi \in \mathcal{F}'_\theta(\mathcal{N}') \tag{10}$$

which says that the Laplace transform maps the convolution product on $\mathcal{F}'_\theta(\mathcal{N}')$ into the usual pointwise product in the algebra of functions $\mathcal{G}_{\theta^*}(\mathcal{N})$. Therefore we may use Theorem 2.1 to define the convolution product between two generalized functions as

$$\Phi * \Psi = \mathcal{L}^{-1}(\mathcal{L}\Phi\mathcal{L}\Psi).$$

This allows us to introduce the convolution exponential of a generalized function. In fact, for every $\Phi \in \mathcal{F}'_\theta(\mathcal{N}')$ we may easily check that $\exp(\mathcal{L}\Phi) \in \mathcal{G}_{e^{\theta^*}}(\mathcal{N})$. Using the inverse Laplace transform and the fact that any Young function θ verifies the property $(\theta^*)^* = \theta$ we obtain that $\mathcal{L}^{-1}(\mathcal{G}_{e^{\theta^*}}(\mathcal{N})) = \mathcal{F}'_{(e^{\theta^*})^*}(\mathcal{N}')$. Now we give the definition of the convolution exponential of $\Phi \in \mathcal{F}'_\theta(\mathcal{N}')$, denoted by $\exp^* \Phi$

$$\exp^* \Phi := \mathcal{L}^{-1}(\exp(\mathcal{L}\Phi)).$$

Notice that $\exp^* \Phi$ is a well defined element in $\mathcal{F}'_{(e^{\theta^*})^*}(\mathcal{N}')$ and therefore the distribution $\exp^* \Phi$ is given in terms of a convergent series

$$\exp^* \Phi = \delta_0 + \sum_{n=1}^{\infty} \frac{1}{n!}\Phi^{*n}, \tag{11}$$

where Φ^{*n} is the convolution of Φ with itself n times, $\Phi^{*0} := \delta_0$ by convention with δ_0 denoting the Dirac distribution at 0. We refer to [1] for more details concerning convolution product on $\mathcal{F}'_\theta(\mathcal{N}')$.

A one parameter generalized stochastic process with values in $\mathcal{F}'_\theta(\mathcal{N}')$ is a family of generalized functions $\{\Phi(t),\ t \geq 0\} \subset \mathcal{F}'_\theta(\mathcal{N}')$. The process $\Phi(t)$ is said to be continuous if the map $t \mapsto \Phi(t)$ is continuous. For a given continuous generalized stochastic process $(X(t))_{t\geq 0}$ we define the generalized stochastic process

$$Y(t,\omega,x) = \int_0^t X(s,\omega,x)\, ds \in \mathcal{F}'_\theta(\mathcal{N}')$$

by

$$\mathcal{L}\left(\int_0^t X(s,\omega,x)ds\right)(\xi,p) := \int_0^t \mathcal{L}X(s,\xi,p)\, ds. \tag{12}$$

The process $Y(t,\omega,x)$ is differentiable and we have $\frac{\partial}{\partial t}Y(t,\omega,x)=X(t,\omega,x)$. The details of the proof can be seen in [10, Proposition 11].

2.3. Convolution inverse of distributions

Let Φ be a fixed element on the distribution space $\mathcal{F}'_\theta(\mathcal{N}')$ and consider the following convolution equation

$$\Phi * \Psi = \delta_0 \,. \tag{13}$$

Applying the Laplace transform to the convolution equation (13) we obtain

$$\hat{\Phi}.\hat{\Psi} = 1 \,.$$

If $\hat{\Phi}(\xi, q) \neq 0$ for every $(\xi, q) \in \mathcal{N}$, then using the division result in the space $\mathcal{G}_{\theta^*}(\mathcal{N})$ (see [6]) we obtain

$$\hat{\Psi} = \frac{1}{\hat{\Phi}} \in \mathcal{G}_{\theta^*}(\mathcal{N}) \,.$$

Moreover, by the Laplace transform isomorphism (see Theorem 2.1), we prove the existence and uniqueness of the solution $\Psi \in \mathcal{F}'_\theta(\mathcal{N}')$ in the equation (13). If we denote this solution Ψ by Φ^{*-1}, we have

$$\mathcal{L}\big(\Phi^{*-1}\big) = \frac{1}{\mathcal{L}(\Phi)} .$$

This division result is also true in the limit case $\theta(x) = (x, x)$; i.e., $\hat{\Phi} \in \mathcal{G}_{\theta^*}(\mathcal{N}) = \mathrm{Hol}_0(\mathcal{N})$.

3. Solution of the n-dimensional convolution equation

We are now ready to solve the Cauchy problem stated in (1) which we recall for the reader convenience, namely

$$\begin{cases} \dfrac{\partial \vec{u}}{\partial t} + (\vec{u} * \vec{\nabla})\vec{u} = \nu \Delta \vec{u} + \vec{f} * \vec{u}, \\ \vec{u}(0, x) = \vec{u}_0(x). \end{cases} \tag{14}$$

The different terms in (14) are as follows: $\vec{u}_0(x) = (u_{0,1}(x), \ldots, u_{0,n}(x))$ is a generalized function; $u_{0,j}(x) \in \mathcal{F}'_\theta(\mathcal{N}')$, $\nu > 0$ a real constant, $x = (x_1, \ldots, x_n) \in \mathbb{R}^n$, $\vec{u} = (u_1, \ldots, u_n)$, $\vec{f} = (f_1, \ldots, f_n)$, with $f_j = f_j(t, x)$,

$u_j \in \mathcal{F}'_\theta(\mathcal{N}')$. We denote

$$\vec{f} * \vec{u} = (f_1 * u_1, \ldots, f_n * u_n)$$

$$\Delta\vec{u} = \begin{pmatrix} \Delta u_1 \\ \vdots \\ \Delta u_n \end{pmatrix}$$

$$(\vec{u} * \nabla)\vec{u} = \begin{pmatrix} \sum_{j=1}^n u_j * \partial_j u_1 \\ \vdots \\ \sum_{j=1}^n u_j * \partial_j u_n \end{pmatrix}, \quad \partial_j := \frac{\partial}{\partial x_j}, \; j = 1, \ldots, n.$$

We are now ready to prove the main result of this paper, namely we obtain the explicit solution of the equation (14), using the tools from Section 2.

Theorem 3.1. *Let $\vec{u}_0(x) = (u_{0,1}(x), \ldots, u_{0,n}(x))$ and $\vec{f} = (f_1, \ldots, f_n)$ be such that $u_{0,k}, f_k \in \mathcal{F}'_x(\mathcal{N}')$, $k = 0, \ldots, n$. Then the solution $\vec{u}(t, \omega, x)$ of the nonlinear equation of convolution type (14) is given explicitly by the following system:*

$$\begin{aligned} u_k(t,x) &= u_{0,k}(x) * e^{* \int_0^t f_k(s)ds} * \gamma_{2\nu t} \\ &\quad * \left(\delta_0 + \sum_{j=1}^n \partial_j u_{0,j} * \int_0^t e^{* \int_0^\tau f_j(s)ds} * \gamma_{2\nu\tau}\, d\tau \right)^{*-1} \end{aligned} \tag{15}$$

with $k = 1, \ldots, n$ and δ_0 is the Dirac measure at point zero and $\gamma_{2\nu t}$ is the Gaussian measure with variance $2\nu t$ on $\mathbb{R}^n$.

Proof. We denote $\partial_t = \frac{\partial}{\partial t}$ such that the n-dimensional equation (14) may be written as

$$\begin{cases} \partial_t u_k + \sum_{j=1}^n u_j * \partial_j u_k = \nu \Delta u_k + f_k * u_k, \\ u_k(0, \omega, x) = u_{0,k}(\omega, x), \end{cases} \tag{16}$$

where $k = 1, \ldots, n$.

We denote by $v_k = v_k(t, \xi, q)$, $g_k = g_k(t, \xi, q)$ and $v_{0,k} = v_{0,k}(\xi, q)$, $\xi \in S(\mathbb{R}, \mathbb{R}^d)$, $q \in \mathbb{R}^n$, the Laplace transforms of the generalized functions $u_k = u_k(t, \omega, x)$, $f_k = f_k(t, \omega, x)$ and initial condition $u_{0,k} = u_{0,k}(\omega, x)$, respectively, for $k = 1, \ldots, n$. Applying the Laplace transform to the system

(16) we obtain

$$\begin{cases} \partial_t v_k - \sum_{j=1}^{n} q_j v_j v_k = \nu q^2 v_k + g_k v_k \\ v_k(0,q) = v_{0,k}(q). \end{cases} \tag{17}$$

Changing the variables, $S_k = \frac{1}{v_k}$, $k = 1, \dots, n$, the system (17) is equivalent to the following:

$$\begin{cases} -\partial_t S_k - \sum_{j=1}^{n} q_j \frac{S_k}{S_j} = (\nu q^2 + g_k) S_k, \\ S_k(0,\xi,q) = S_{0,k}(\xi,q). \end{cases} \tag{18}$$

We denote $S_1 \cdots \check{S}_k \cdots S_n = S_1 \cdots S_{k-1} S_{k+1} \cdots S_n$, for $k = 1, \dots, n$. If we multiply the first equation of the system (18) by $S_1 \cdots \check{S}_k \cdots S_n$, we deduce

$$(-\partial_t S_k) S_1 \cdots \check{S}_k \cdots S_n - \sum_{j=1}^{n} q_j S_1 \cdots \check{S}_j \cdots S_n = (\nu q^2 + g_k) S_1 \cdots S_n. \tag{19}$$

Let us denote such equation by (E_k), $k = 1, \dots, n$. If we fix k, then the difference $(E_1) - (E_k)$ becomes

$$\check{S}_1 S_2 \cdots \check{S}_k \cdots S_n (-S_k \partial_t S_1 + S_1 \partial_t S_k) = (g_1 - g_k) S_1 \cdots S_n. \tag{20}$$

After simplification, we divide by $S_1 S_k$ and obtain the equation

$$\frac{\partial_t S_k}{S_k} - \frac{\partial_t S_1}{S_1} = g_1 - g_k \tag{21}$$

which can be integrated. The solution is

$$S_k = \frac{S_{0,k}}{S_{0,1}} S_1 \, e^{\int_0^t (g_1(s) - g_k(s)) ds}, \quad k = 1, \dots, n, \tag{22}$$

where for simplification of notation $g_j(s) = g_j(s,\xi,q)$, $j = 1, \dots, n$. Then we have the relation

$$\frac{S_k}{S_j} = \frac{S_{0,k}}{S_{0,j}} e^{\int_0^t (g_j(s) - g_k(s)) ds}, \quad k, j = 1, \dots, n. \tag{23}$$

Introducing the expression of $\frac{S_k}{S_j}$ in (18) we deduce the following linear system of equations

$$\partial_t S_k + \sum_{j=1}^{n} q_j \frac{S_{0,k}}{S_{0,j}} e^{\int_0^t (g_j(s) - g_k(s)) ds} = -(\nu q^2 + g_k) S_k, \quad k = 1, \dots, n. \tag{24}$$

For any fixed k, the solution of the homogeneous equation is given by

$$S_k(t,q) = \lambda e^{-\int_0^t (\nu q^2 + g_k(s))ds}, \tag{25}$$

where λ is a constant. Then the solution of (24) is given by the method of variation of constants as

$$\begin{aligned} S_k(t,\xi,q) &= \lambda e^{-\int_0^t (\nu q^2 + g_k(s))ds} \\ &\quad - e^{-\int_0^t (\nu q^2 + g_k(s))ds} \sum_{j=1}^n q_j \frac{S_{0,k}}{S_{0,j}} \int_0^t e^{\int_0^\tau (g_j(s)+\nu q^2)ds}\, d\tau, \end{aligned} \tag{26}$$

where the constant λ is determined by the initial conditions; $S_k(0,\xi,q) = S_{0,k} = \lambda$. Then (26) may be written as

$$\begin{aligned} S_k(t,\xi,q) &= S_{0,k}\, e^{-\int_0^t g_k(s)ds}\, e^{-\nu q^2 t} \\ &\quad \left(1 - \sum_{j=1}^n \frac{q_j}{S_{0,j}} \int_0^t e^{\int_0^\tau (g_j(s)+\nu q^2)ds}\, d\tau\right). \end{aligned} \tag{27}$$

Since $S_k(t,\xi,q) = \dfrac{1}{v_k(t,\xi,q)}$, we obtain

$$\begin{aligned} v_k(t,\xi,q) &= v_{0,k}\, e^{\int_0^t g_k(s)ds}\, e^{\nu q^2 t} \\ &\quad \left(1 - \sum_{j=1}^n q_j v_{0,j} \int_0^t e^{\int_0^\tau (g_j(s)+\nu q^2)ds}\, d\tau\right)^{-1}. \end{aligned} \tag{28}$$

In fact, it is easy to show that for every $t \geq 0$, the function

$$Y(t,q,\xi) = 1 - \sum_{j=1}^n q_j v_{0,j}(\xi,q) \int_0^t e^{\int_0^\tau (g_j(s,\xi,q)+\nu q^2)ds}\, d\tau$$

belongs to the space $\mathrm{Hol}_0(\mathcal{N})$ and satisfies $Y(t,0,0) = 1 \neq 0$. Then there exists $\mathcal{U}$ a neighborhood of $(0,0)$ of $\mathcal{N}$, such that $Y(t,q,\xi) \neq 0$ for every $(\xi,q) \in \mathcal{U}$. Therefore

$$\frac{1}{Y(t,q,\xi)} \in \mathrm{Hol}_0(\mathcal{N})$$

which implies that

$$v_k(t,q,\xi) \in \mathrm{Hol}_0(\mathcal{N}).$$

Finally, to obtain the solution of the equation (14) we use the following equalities:

$$\mathcal{L}\left\{e^{*\int_0^t f_k(s)ds}\right\} = e^{\int_0^t g_k(s)ds}$$

$$\mathcal{L}\left\{\int_0^t e^{*\int_0^\tau f_k(s)ds} * \gamma_{2\nu\tau})\, d\tau\right\} = \int_0^t e^{\int_0^\tau g_k(s)ds} e^{\nu q^2 \tau}\, d\tau$$

and then $u_k(t,x)$, $k = 1, \ldots, n$ is given by the Laplace inverse transform according to the theorem 2.1, as in (15). □

Corollary 3.1. *If the potential $\vec{f}$ in the equation (14) does not depend on the time variable t, i.e., $\vec{f} = \vec{f}(x)$, then the solution is given by*

$$u_k(t,x) = u_{0,k}(x) * e^{*tf_k} * \gamma_{2\nu t} * \left(\delta_0 + \sum_{j=1}^n \partial_j u_{0,j} * \int_0^t e^{*\tau f_j} * \gamma_{2\nu\tau}\, d\tau\right)^{*-1}$$

with $k = 1, \ldots, n$. In particular, if $f = 0$ the solution has the from

$$u_k(t,x) = u_{0,k}(x) * \gamma_{2\nu t} * \left(\delta_0 + \nabla \cdot u_0 * \int_0^t \gamma_{2\nu\tau} d\tau\right)^{*-1}$$

with $k = 1, \ldots, n$ and $\nabla\cdot$ represents the divergence operator.

Acknowledgment

We thank Martin Grothaus for useful discussions. Financial support by GRICES, Portugal/Tunisia, 2004 and FCT, POCTI – Programa Operacional Ciência, Tecnologia e Inovação, FEDER are gratefully acknowledged.

References

1. M. Ben Chrouda, M. El Oued, H. Ouerdiane, "Convolution calculus and applications to stochastic differential equations", *Soochow J. Math.* **28**(4), 375–388 (2002).
2. Yu. M. Berezansky, Yu. G. Kondratiev, *Spectral Methods in Infinite-Dimensional Analysis*, volume 1, Kluwer Academic Publishers, Dordrecht, 1995.
3. J. M. Burgers, *The Nonlinear Diffusion Equation.* D. Reidel Publishing Company, Dordrecht, Holland, 1974.
4. J. D. Callen, C. C. Hegana, E. D. Held, T. A. Gianakon, S. E. Kruger, C. R. Sovinec, "Nonlocal closures for plasma fluid simulations", *Phys. Plasmas* **11**(5), 2419–2426, (2004).

5. Weinan E, K. Khanin, A. Mazel, Ya. Sinai, Invariant measures for Burgers equation with stochastic forcing, *Ann. of Math. (2)* **151**(3), 877–960 (2000).
6. R. Gannoun, R. Hachaichi, P. Kree, H. Ouerdiane, "Division de fonctions holomorphes a croissance θ-exponentielle", Technical Report E 00-01-04, BiBoS University of Bielefeld, 2000.
7. R. Gannoun, R. Hachaichi, H. Ouerdiane, A. Rezgui, "Un théoréme de dualité entre espaces de fonctions holomorphes à croissance exponentielle", *J. Funct. Anal.* **171**(1), 1–14 (2000).
8. T. Hida, H. H. Kuo, J. Potthoff, L. Streit, *White Noise. An Infinite Dimensional Calculus.* Kluwer Academic Publishers, Dordrecht, 1993.
9. H. Ouerdiane, "Infinite dimensional entire functions and applications to stochastic differential equations", *Not. S. Afr. Math. Soc.* **35**(1), 23–45 (2004).
10. H. Ouerdiane, J. Silva, "On the heat equation with positive generalized stochastic process potential", *Methods Funct. Anal. Topology* **10**(3), 54–63 (2004).
11. H. Ouerdiane, J. L. Silva, "Generalized Feymann-Kac formula with stochastic potential", *Infin. Dimens. Anal. Quantum Probab. Relat. Top.* **5**(2), 1–13 (2002).
12. A. J. Roberts, "Planform evolution in convection — an embedded centre manifold", *J. Austral. Math. Soc. Ser. B* **34**(2), 174–198 (1992).
13. A. J. Roberts, "The swift-Hohenberg equation requires non-local modifications to model spatial pattern evolution of physical problems", technical report, `arXiv:patt-sol/9412002` (2002).
14. P. Ván, "Weakly nonlocal continuum theories of granular media: restrictions from the second law", *International Journal of Solids and Structures* **41**(21), 5921–5927 (2004).
15. P. Ván, T. Fülöp, "Weakly nonlocal fluid mechanics — the Schrödinger equation", technical report, `arXiv:quant-ph/0304062` (2004).

F. Cipriano
GFM e Departamento de Matemática FCT-UNL,
Av. Prof. Gama Pinto 2, 1649-003, Lisboa, Portugal
`cipriano@gfm.cii.fc.ul.pt`

H. Ouerdiane
Département de Mathématiques, Faculté des Sciences de Tunis,
1060 Tunis, Tunisia
`habib.ouerdiane@fst.rnu.tn`

J. L. Silva
University of Madeira, CCM, 9000-390 Funchal, Portugal
`luis@uma.pt`

R. Vilela Mendes
CMAF, Complexo Interdisciplinar, Universidade de Lisboa,
Av. Prof. Gama Pinto 2, 1649-003 Lisboa, Portugal
`vilela@cii.fc.ul.pt`

VARIATIONAL PRINCIPLE FOR DIFFUSIONS ON THE DIFFEOMORPHISM GROUP WITH THE H^2 METRIC

FERNANDA CIPRIANO (*GFM-UL / FCT-UNL, Lisboa*),
ANA BELA CRUZEIRO (*GFM-UL / IST, Lisboa*)

We describe a stochastic variational principle associated to the H^2 metric on the group of diffeomorphisms of the torus as well as the corresponding dynamics. This variational principle corresponds to a minimal entropy principle.

1. Introduction

In his celebrated paper [1], Arnold has shown that the Euler equation for incompressible fluids can be characterized as the geodesic equation in the group of volume preserving diffeomorphisms with respect to the weak L^2 invariant metric. His work, developed by Ebin and Marsden [6], introduced Lagrangian formalism in hydrodynamics and showed the relevance of infinite dimensional geometry in this area.

Considering the H^1 right invariant metric, the geometry of the volume preserving diffeomorphism group and the corresponding equations of motion were studied in [13], inspired by a hydrodynamical model which had been previously introduced by Holm, Marsden and Ratiu [9].

In [3], and in the two dimensional periodic case, we proposed a generalization of Arnold's result that consists in obtaining Navier-Stokes equations as critical points of a stochastic variational action defined in the space of measure preserving homeomorphisms with respect to the L^2 metric. It is an extension of the deterministic variational principle in the sense that the Laplacian term of Navier-Stokes is associated to a random perturbation of the deterministic Euler system. This approach found its inspiration in the work [12]. On the other hand it relies on the possibility of solving some stochastic differential equations on the space of homeomorphisms of the torus and for this uses the recent developments on the subject ([11], [7]).

Other variational characterizations of Navier-Stokes equations can be found in the literature and follow different points of view. Without trying to be exhaustive, we refer to [10] and to [8].

Here we adopt a similar point of view of [3] but consider a H^2 metric. We describe the stochastic variational approach of the associated dynamics.

The higher regularity of the metric allows to reinterpret the variational principle for the corresponding energy functional as a minimal entropy principle. The same methods can actually be applied to $H^{1+\epsilon}$ with any $\epsilon > 0$; then all the corresponding results hold true, except the partial differential equation coming from the variational principle (Theorem 2), which has to be modified accordingly.

2. Diffusions on the homeomorphism group of the torus

We denote by $\mathbb{T}$ the two-dimensional torus, $d\theta$ its volume element measure.

$$G^s = \{g : \mathbb{T} \to \mathbb{T} \ \text{ bijection}, \ g, g^{-1} \in H^s\}$$

is the group of homeomorphisms of the torus belonging to the Sobolev class H^s and G^s_V the subgroup of those which preserve the volume. For $g \in H^s$, G^s is locally diffeomorphic to the Hilbert space

$$H^s_g = \{Z \ \text{ vector field on } \mathbb{T} : \pi o Z = g\},$$

where π is the natural projection from the tangent space of $\mathbb{T}$ to $\mathbb{T}$. More precisely, a local chart is defined by $\omega_{exp} : H^s_g \to G^s$, $\omega_{exp}(Z) = exp\, oZ$, where exp is the Riemannian exponential map of the metric on $\mathbb{T}$.

For the case of volume preserving diffeomorphisms, the Lie algebra of G^s_V, denoted here by $\mathcal{G}^s_V$, is the space of H^s Sobolev vector fields on the torus with zero divergence.

The Euler equation corresponds to a flow on G^0_V which is critical for the action functional defined by the square of the right invariant metric L^2 and defined on G^0_V, namely,

$$S[g](t) = \frac{1}{2} \int_0^T ||\dot{g}(t)||^2_{L^2}\, dt\,.$$

We consider the complete system of $\mathcal{G}^0_V$ given by the trigonometric polynomials,

$$A_k = \frac{1}{|k|}(k_2 \cos(k.\theta), -k_1 \cos(k.\theta)),$$

$$B_k = \frac{1}{|k|}(k_2 \sin(k.\theta), -k_1 \sin(k.\theta)),$$

where $k = (k_1, k_2) \in \mathbb{Z}^2$, $k \neq 0$. Notice that we are considering vector fields with mean zero, an assumption that is not restrictive from the point of view of the dynamics we study.

We follow the methodology of [11] and [7] in order to construct a Brownian motion that takes values in the group of homeomorphisms (for more details about this construction in the two dimensional case, see [3]). We consider a sequence of $\mathbb{R}^2$-valued independent Brownian motions $x_k = (x_k^1, x_k^2)$ which are adapted to the usual increasing filtration (generated by events before time t) and define

$$dx(t) = \sum_{k\neq 0} \frac{1}{|k|^3} \left[A_k\, dx_k^1(t) + B_k\, dx_k^2(t)\right].$$

This sum can be shown to converge uniformly in $[0,T] \times \mathbb{T}$ and to have quadratic variation

$$E\,|x(t)^i(\theta) - x(t)^i(\theta')|^2 \leq C\,|\theta - \theta'|^2 \log \frac{1}{|\theta - \theta'|}.$$

Consider the stochastic differential equation

$$dg(t) = (odx(t))(g(t)), \quad g(0) = \mathrm{id}, \tag{1}$$

which is a condensed form of writing

$$dg^1(t) = \sum_{k\neq 0} \frac{1}{|k|^3} \left[A_k^1(g(t)) \circ dx_k^1(t) + B_k^1(g(t)) \circ dx_k^2(t)\right],$$

$$dg^2(t) = \sum_{k\neq 0} \frac{1}{|k|^3} \left[A_k^2(g(t)) \circ dx_k^1(t) + B_k^2(g(t)) \circ dx_k^2(t)\right].$$

Then the process $g(t)$ is well defined and lives in G_V^0.

An important observation (cf. [3]) is that the Itô contraction vanishes and therefore Stratonovich stochastic differentiation in the definition of $g(t)$ coincides with Itô differentiation. This can be checked by direct computation.

Define the generator of the process $g(t)$ by

$$\mathcal{L}(F)(g)(\theta) = \lim_{t\to 0} \frac{1}{t} E\Big(F(g(t))(\theta) - F(g)(\theta)\Big).$$

Then, using Itô calculus, we can prove the following

Theorem 1. *([3])*
When considered on functions on the torus, namely when $F(g)(\theta) = f(g(\theta))$ for f smooth, the generator of the process $g(t)$ applied to F coincides with the usual Laplacian of f multiplied by the constant $c = 2\sum_{k\neq 0} k_1^2/|k|^6$, $\mathcal{L}F(g)(\theta) = c\Delta f(g(\theta))$.

Let $u_t(g)$ be a vector field on G_V^2, continuous in t and satisfying

$$\sup_{t,g} ||u_t(g)||_{H^2} < +\infty\,. \tag{2}$$

We can solve, proceeding as in [7], the following stochastic differential equation

$$dg^u(t) = \left(\sqrt{\frac{\nu}{c}}\, dx(t) + u_t\right)(g^u(t)), \qquad g^u(0) = \mathrm{id}, \tag{3}$$

weakly via a Girsanov transformation. The law of this process at each time t has for density with respect to the law of $g(t)$ the martingale (with respect to the future filtration)

$$M_t = \exp\Bigg\{ \sum_k \Bigg[|k|^2 \sqrt{\frac{c}{\nu}} \int_0^t \Big(u^{A_k}(g(s))\, dx_k^1(s) + u^{B_k}(g(s))\, dx_k^2(s) \Big) - \frac{c|k|^4}{2\nu} \Big((u^{A_k}(g(s)))^2 + (u^{B_k}(g(s)))^2 \Big)\, ds \Bigg] \Bigg\},$$

where we have used the notation:

$u^{A_k} = \langle u, A_k \rangle_{L^2}$, $u^{B_k} = \langle u, B_k \rangle_{L^2}$, if $k_1 > 0$ and k_2 arbitrary;
$u^{A_k} = u^{B_k} = 0$, otherwise.

3. The variational principle and the associated dynamics

We denote by $\mathcal{S}$ the set of continuous semimartingales $\xi(t)$, $t \in [0,T]$ with initial condition $g(0) = \mathrm{id}$ and taking values in G_V^0.

On $\mathcal{S}$ we consider the action functional

$$S[\xi] = \frac{1}{2} E \int_0^T ||D\xi(t)||_{H^2}^2 dt - \frac{1}{2} E ||D\xi(T)||_{H^2}^2, \tag{4}$$

where D is the mean derivation operator

$$DF(\xi(t)) = \lim_{\text{a.s.},\delta \to 0} \frac{1}{\delta} E^{\mathcal{F}_t} \big[F(t+\delta, \xi(t+\delta)) - F(t, \xi(t)) \big].$$

The variations will be defined as follows. For $v \in C^1([0,T]; \mathcal{G}_V^\infty)$ with $v(0) = 0$ we consider the G_V^∞-valued deterministic path $e_t(v)$ which is the solution of the ordinary differential equation

$$\frac{d}{dt} e_t(v) = \dot{v}(t, e_t(v)), \qquad e_0 = \mathrm{id}\,.$$

We compute the left derivative of the action functional S at a process ξ, this left derivative being defined as

$$D_{L,v}S[\xi] = \frac{d}{d\delta} S[e(\delta v)\circ\xi]\Big|_{\delta=0}.$$

For $\xi(t) = g^u(t)$ with u as in (2), we have

$$\frac{d}{d\delta}(e_t(\delta v)\circ g^u(t))\Big|_{\delta=0} = v_t \circ g^u(t).$$

Using Taylor expansion, the perturbation of the process $g^u(t)$ in the direction $e_t(v)$ can be written as

$$e_t(\delta v)\circ g^u(t) = g^u(t) + \delta v_t(g^u(t)) + o(\delta).$$

We have

$$S[e_t(\delta v)\circ g^u(t)] = \frac{1}{2}E\int_0^T \left\|u(g^u(t)) + \mathcal{L}^u(v))(g^u(t)) + o(\delta)\right\|_{H^2}^2 dt,$$

where $\mathcal{L}^u$ denotes the infinitesimal generator of the process g^u, namely

$$\mathcal{L}^u(v)^i = \frac{\partial v^i}{\partial t} + u.\nabla v^i + \nu\Delta v^i, \quad i = 1, 2.$$

Therefore

$$\begin{aligned} D_{L,v}S[g^u] &= \frac{d}{d\delta}S[e(\delta v)\circ g^u)]\Big|_{\delta=0} \\ &= E\int_0^T\int_{\mathbb{T}}\Big(\Delta^2 u, \frac{\partial v}{\partial t} + u.\nabla v + \nu\Delta v\Big)(g^u(t))\, d\theta\, dt. \end{aligned}$$

Using the fact that g^u preserves the measure $d\theta$ and integrating by parts,

$$D_{L,v}S[g^u] = \int_0^T\int_{\mathbb{T}}\Big(-\frac{\partial}{\partial t}\Delta^2 u_t + \nu\Delta^3 u_t - (u_t.\nabla)\Delta^2 u_t\Big) v_t\, dt.$$

We have deduced the following

Theorem 2. *The process $g^u(t)$, with $u_t \in H^2$ as in (2), is a critical value of the action functional S in (4) if and only if u satisfies the partial differential equation*

$$\frac{\partial}{\partial t}\Delta^2 u = \nu\Delta^3 u - (u.\nabla)\Delta^2 u + \nabla p, \qquad u_T \in H^2, \qquad \operatorname{div} u = 0,$$

for some function p.

4. The minimal entropy principle

On a countably generated measurable space $(E, \mathcal{B})$, and for σ and μ two probability measures on E, the relative entropy of σ with respect to μ is defined as

$$h_{\mathcal{B}}(\sigma;\mu) = \int_E \frac{d\sigma}{d\mu} \log \frac{d\sigma}{d\mu}\, d\mu$$

if σ is absolutely continuous with respect to μ and $+\infty$ otherwise.

On $\Omega = C([0,1]; G_V^0, \mathcal{F}_0^T)$ we consider the canonical realization of the Brownian motion $g(t)$ defined in paragraph 2. Its law on Ω will be denoted by $\mathbb{P}$.

For a probability measure $\mathbb{Q}$ on Ω denote by $\mathbb{Q}_t$ the marginal law of $g(t)$ under $\mathbb{Q}$. Assume $\mathbb{Q}_0 = \delta_e$ is the Dirac mass at the identity and $\mathbb{Q}_T = \mu_T$. If $\mathbb{Q}$ is absolutely continuous with respect to $\mathbb{P}$, let M_t be the density martingale

$$M_t = \left.\frac{d\mathbb{Q}}{d\mathbb{P}}\right|_{\mathcal{F}_0^t}, \quad t \in [0,T].$$

Then the coordinate process satisfies the equation

$$dg(t) = (d\tilde{x}(t) + u_t)(g(t)),$$

where $\tilde{x}(t)$ is a $\mathbb{Q}$-Brownian motion and

$$u(t, g(t)) = \lim_{\delta\to 0} \frac{1}{\delta} E^{\mathbb{Q}}((g(t+\delta) - g(t))|\mathcal{F}_0^t) = Dg(t).$$

It can be proved (cf. [4]) that the regularized velocity term of the action functional, namely

$$S(\mathbb{Q}) = \frac{1}{2} E^{\mathbb{Q}} \int_0^T \|Dg(t)\|_{H^2}\, dt$$

coincides with the relative entropy, i.e.,

$$S(\mathbb{Q}) = h_{\mathcal{F}_0^t}(\mathbb{Q}, \mathbb{P}_{\delta_e}).$$

An application of Csiszar's minimization theorem ([5]) provides an unique probability measure $\mathbb{Q}$ that attains the infimum of $S(\mathbb{Q})$ for $\mathbb{Q}$ probability measures on Ω with $\mathbb{Q}_0 = \delta_e$ and $\mathbb{Q}_T = \mu_T$, if this infimum is not equal to infinity, which is true by Theorem 1. Moreover the corresponding coordinate process is Markov.

Other properties of this process could be studied, notably the multiplicative nature of its law (cf. [4]).

Acknowledgements

Research of both authors has been carried out in the framework of the Project POCI/MAT/55977/2004 (co-financed by FCT/OE and FEDER through POCI2010).

References

1. V. Arnold, "Sur la géométrie différentielle des groupes de Lie de dimension infinie et ses applications a l'hydrodynamique des fluides parfaits", *Ann. Inst. Fourier* **16**, 319–361 (1966).
2. P. Cattiaux, C. Leonard, "Minimization of the Kullback information for some Markov processes", in *Seminaire de Probabilites XXX*, *Lect. Notes in Mathematics* 1626, Springer, 1996, pp. 288–311.
3. F. Cipriano, A. B. Cruzeiro, "Navier-Stokes equation and diffusions on the group of homeomorphisms of the torus", preprint.
4. A. B. Cruzeiro, W. Liming, J. C. Zambrini, "Bernstein processe associated with a Markov process", in *Stochastic Analysis and Mathematical Physics, Trends in Mathematics*, ed. R. Rebolledo, Birkhäuser, 2000.
5. I. Csiszar, "I-divergence geometry of probability distributions and minimization problems", *Ann. Probab.* **3**(1), 146 (1975).
6. D. Ebin, J. Marsden, "Groups of diffeomorphisms and the motion of an incompressible fluid", *Ann. of Math.* **92**, 22–46 (1970).
7. S. Fang, "Solving stochastic differential equations on Homo ($S1$)", *J. Funct. Anal.* **216**, 22–46 (2004).
8. D. A. Gomes, "A variational formulation for the Navier-Stokes equation", *Comm. Mathem. Physics* **257**(1), 227–234 (2005).
9. D. D. Holm, J. Marsden, T. S. Ratiu, "Euler-Poincaré models of ideal fluids with nonlinear dispersion", *Phys. Rev. Letters* **80**, 4273–4277 (1998).
10. A. Inoue, T. Funaki, "A new derivation of the Navier-Stokes equation", *Comm. Mathem. Physics* **65**, 83–90 (1979).
11. P. Malliavin, "The canonic diffusion above the diffeomorphism group of the circle", *C. R. Acad. Sci. Paris* **329**, 325–329 (1999).
12. T. Nakagomi, K. Yasue, J. C. Zambrini, "Stochastic variational derivations of the Navier-Stokes equation", *Letters in Math. Physics* **160**, 337–365 (1981).
13. S. Shkoller, "Geometry and curvature of diffeomorphism groups with H^1 metric and mean hydrodynamics", *J. Funct. Anal.* **5**, 545–552 (1981).

Fernanda Cipriano
Grupo de Física-Matemática UL and Dep. de Matemática FCT-UNL,
Av. Prof. Gama Pinto 2, 1649-003 Lisboa, Portugal
`cipriano@cii.fc.ul.pt`

Ana Bela Cruzeiro
Grupo de Física-Matemática UL and Dep. Matemática IST,
Av. Rovisco Pais, 1049-001 Lisboa, Portugal
`abcruz@math.ist.utl.pt`

ON A VARIATIONAL PRINCIPLE FOR THE NAVIER-STOKES EQUATION

DIOGO AGUIAR GOMES (*IST, Lisboa*)

In this paper we study the variational principle for the Navier-Stokes equation described in [14], and clarify the role of boundary conditions. We show that in certain special cases this variational principle gives rise to new models for fluid equations.

1. Introduction

Several authors, for instance [19], [18], [3], [2], [8], have studied representation formulas for solutions of the Navier-Stokes equation using probabilistic methods, and, in fact, the idea of using random maps instead of deterministic ones can be traced back to Chorin [6] and Peskin [17]. In a completely different setting, Arnold [1], and Ebin and Marsden [12], proved a variational principle for the solutions of Euler equation. It is of course natural to ask whether there is a probabilistic analog for the Navier-Stokes equation. In fact, there is at least one, non-probabilistic variational principle for the Navier-Stokes equation[13]. However, recently stochastic variational principles for the Navier-Stokes equation have drawn attention and at least three different ones have been proposed by several authors: [15], [5], and [14]. This paper builds upon the last variational principle and is a contribution to its study. This variational principle can be stated as follows:

Theorem 1.1. *Let B_t be a Brownian motion defined on a probability space Ω, and let E denote the expected value on Ω. Let $\Theta(X,x) : \mathbb{R}^n \times \mathbb{R}^n \to \mathbb{R}^n$ be a smooth function with suitable growth conditions. Suppose that for each $\omega \in \Omega$, (u, ϕ^ω) is a critical point, u smooth in space-time, and ϕ smooth in space and C^1 in time, of*

$$\hat{S} = \int_0^T \int_{\mathbb{R}^n} \frac{|u|^2}{2} + E \int_{\mathbb{R}^n} \Theta(\phi^\omega(x,T), x)$$

under the constraint $\operatorname{div} u = 0$, *and*

$$\frac{\partial \phi^\omega}{\partial t} = u \circ B_t \circ \phi^\omega, \tag{1}$$

and $\phi^\omega(x,0)=x$. Assume further that the function Π^ω which satisfies the differential equation

$$\frac{\partial \Pi_i^\omega}{\partial t}+\Pi_j^\omega D_i u_j \circ B_t \circ \phi^\omega = 0, \tag{2}$$

with the terminal condition

$$\Pi^\omega(x,T)=D_X\Theta(\phi^\omega(x,T),x),$$

is, at $t=0$, non-random. Then u is a solution to the Navier-Stokes equation

$$u_t+u\cdot\nabla u+\nabla p=\frac{1}{2}\Delta u,$$

where p is the scalar pressure field.

One can relate some of the objects that arise in this last theorem to the works by P. Constantin in [9], [10], [11]. In fact, consider a critical point (u,ϕ^ω) as in the previous theorem, and define

$$A=E\left[(\phi^\omega)^{-1}\circ B_t^{-1}\right].$$

The next proposition shows that the vector A satisfies exactly the advection-diffusion equation as in [10].

Proposition 1.1. *The vector A satisfies*

$$\frac{\partial A}{\partial t}+(u\cdot\nabla)A-\frac{1}{2}\Delta A=0,$$

with $A(x,0)=x$. Note that this does not depend on u being a solution to the Navier-Stokes equation.

In this paper we study a variation of theorem 1.1, with a special terminal cost which enable to remove the probability from the variational principle. However, the vector Π will not be in general progressively measurable, hence the critical points may not be solutions to the Navier-Stokes equation. This gives rise to a new model for incompressible viscous fluids and we study some of its properties.

2. A non-probabilistic variational principle

Consider the case of a terminal cost which depends only on the active transport vector A, and given deterministic data. For instance

$$\frac{1}{2}\int_{\mathbb{R}^n}|A(x,T)-\psi(x,T)|^2.$$

Then, we may replace the variable ϕ^ω by the variable A, and so we may consider the problem

$$\min \int_0^T \int_{\mathbb{R}^n} \frac{|u|^2}{2} + \frac{1}{2} \int_{\mathbb{R}^n} |A(x,T) - \psi(x)|^2, \tag{3}$$

under the constraints

$$\operatorname{div} u = 0, \qquad \partial_t A + (u \cdot \nabla) A - \frac{1}{2} \Delta A = 0. \tag{4}$$

Theorem 2.1. *Let (u, A) be a critical point of the functional (3) under the constraints (4). Let ζ be the solution of*

$$\partial_t \zeta_i - \partial_j (\zeta_i u_j) + \frac{1}{2} \Delta \zeta_i = 0, \tag{5}$$

with the terminal condition

$$\zeta(x,T) = A(x,T) - \psi(x).$$

Then $u = \mathbb{P}m$, where $m_i = \zeta_j \partial_i A_j$ satisfies the modified Navier-Stokes equation in magnetization form:

$$\partial_t m_i + u_j D_j m_i + m_j D_i u_j = \frac{1}{2} \Delta m_i - \partial_k \left((\partial_k \zeta_j)(\partial_i A_j) \right). \tag{6}$$

Remark. (6) should be compared with the corresponding identity for the Navier-Stokes equation, see (8) in the appendix.

Remark. The functional (3) has as domain the velocity fields $u \in L^2$, for which the corresponding weak solution of the active transport equation satisfies $A(x,T) - \psi(x) \in L^2(\mathbb{R}^n)$. However, the theorem, as stated, is only valid for smooth critical points.

Remark. The Leray projection $\mathbb{P}m$ is defined as the L^2 projection of the vector field m into the divergence free vector fields in $\mathbb{R}^n$.

Proof. We will use the variable ζ as a Lagrange multiplier to make the computation of the Euler-Lagrange equation easier. Thus consider the equivalent problem of minimizing

$$\int_0^T \int_{\mathbb{R}^n} \frac{|u|^2}{2} + \left\langle \zeta, \partial_t A + (u \cdot \nabla) A - \frac{1}{2} \Delta A \right\rangle + \frac{1}{2} \int_{\mathbb{R}^n} |A(x,T) - \psi(x)|^2$$

under the same constraints. By making variations $A+\epsilon\delta A$ and $u+\epsilon\delta u$ we obtain

$$\int_0^T\int_{\mathbb{R}^n}\langle u,\delta u\rangle+\langle\zeta,(\delta u\cdot\nabla)A\rangle=0,$$

which implies $u=\mathbb{P}m$ with $m_i=\zeta_j\partial_iA_j$, and

$$\int_0^T\int_{\mathbb{R}^n}\left\langle\zeta,\partial_t\delta A+(u\cdot\nabla)\delta A-\frac{1}{2}\Delta\delta A\right\rangle \\ +\int_{\mathbb{R}^n}\langle A(x,T)-\psi(x),\delta A(x,T)\rangle=0,$$

which shows that

$$\zeta(x,T)=A(x,T)-\psi(x),$$

after integration by parts (in space and time) and using the definition of ζ.

Finally, we must show that m satisfies (6). To this end, observe that

$$\begin{aligned}\partial_t(\zeta_j\partial_iA_j)&=\partial_t\zeta_j\partial_iA_j+\zeta_j\partial_i\left(u_k\partial_kA_j+\frac{1}{2}\Delta A_j\right)\\&=\partial_k(\zeta_ju_k)\partial_iA_j+\zeta_j\partial_i\left(u_k\partial_kA_j\right)+\frac{1}{2}\left(\zeta_j\partial_i\Delta A_j-\Delta\zeta_j\partial_iA_j\right)\\&=u_k\partial_km_i+m_k\partial_iu_k\\&\quad+\frac{1}{2}\left(\zeta_j\partial_i\Delta A_j-\Delta\zeta_j\partial_iA_j\right).\end{aligned}$$

To finish the proof, we just have to observe that

$$\begin{aligned}\frac{1}{2}\left(\zeta_j\partial_i\Delta A_j-\Delta\zeta_j\partial_iA_j\right)&=\frac{1}{2}\Delta m_i-\left(\Delta\zeta_j\partial_iA_j+\partial_k\zeta_j\partial_i\partial_kA_j\right)\\&=\frac{1}{2}\Delta m_i-\partial_k\left((\partial_k\zeta_j)(\partial_iA_j)\right).\end{aligned}$$

□

By applying the Leray projection to (6) we obtain that u solves

$$\partial_tu+u\cdot\nabla u=\frac{1}{2}\Delta u+\nabla p-\mathbb{P}(\partial_k\left((\partial_k\zeta_j)(\partial_iA_j)\right)).$$

Finally, we should observe that in the case of no-viscosity, the proof of this theorem is a very short and elementary proof of the well known result by Arnold [1], and Ebin and Marsden [12], for the solutions of Euler's equation, as using the active vector A which is the inverse of the flow map instead of the flow map simplifies a lot the computations.

3. Conservation of Energy

Whereas the original variational problem in [14] does involve some explicit time dependence through the Brownian motion, the modified variational principle we study in this problem does not involve time explicitly. Therefore, by Noether's theorem, there should be a conserved quantity, the energy, corresponding to this symmetry. This is exactly the content of the next theorem. There should be also further symmetries, for instance associated with translation or rotation invariance, although we do not investigate this issue here, also note that this issue was not addressed yet for the original variational problem, although it is not clear that it will produce new identities.

Theorem 3.1 (Energy conservation). *The energy*

$$\int_{\mathbb{R}^n} \frac{|u|^2}{2} - \langle \partial_t \zeta, A \rangle$$

is a constant of motion for the modified Navier-Stokes equation (6).

Proof. Although this conservation law can be derived directly, it is quite instructive to do the proof following the steps of Noether's theorem. We have

$$\begin{aligned}\int_h^{T+h} \int_{\mathbb{R}^n} & \frac{|u(x,t-h)|^2}{2} \\ & + \langle \zeta(x,t-h), \partial_t A(x,t-h) + (u(x,t-h) \cdot \nabla) A(x,t-h) \\ & - \frac{1}{2} \Delta A(x,t-h) \rangle\end{aligned}$$

is independent of h. Therefore, if we differentiate in h at $h=0$ we obtain

$$\begin{aligned}\frac{|u(x,T)|^2}{2} - \frac{|u(x,0)|^2}{2} = & \int_0^T \int_{\mathbb{R}^n} \langle u, \partial_t u \rangle \\ & + \left\langle \zeta, \partial_{tt}^2 A + \partial_t u \cdot \nabla A + u \cdot \nabla \partial_t A - \frac{1}{2} \Delta \partial_t A \right\rangle.\end{aligned}$$

Now observe that

$$\int_0^T \int_{\mathbb{R}^n} \langle u, \partial_t u \rangle + \langle \zeta, \partial_t u \cdot \nabla A \rangle = 0.$$

Furthermore, integrating by parts, we have

$$\int_0^T \int_{\mathbb{R}^n} \left\langle \zeta, \partial_{tt}^2 A + u \cdot \nabla \partial_t A - \frac{1}{2}\Delta \partial_t A \right\rangle$$
$$= \int_{\mathbb{R}^n} \langle \zeta(x,T), \partial_t A(x,T) \rangle - \langle \zeta(x,0), \partial_t A(x,0) \rangle.$$

Since $\partial_t A = -u \cdot \nabla A - \frac{1}{2}\Delta A$ we conclude that

$$\int_{\mathbb{R}^n} \frac{|u|^2}{2} - \left\langle \zeta, u \cdot \nabla A + \frac{1}{2}\Delta A \right\rangle$$

is a constant of motion. Taking into account equation (5) we finish the proof. □

Appendix. Navier-Stokes equation in magnetization variables

In this appendix we review the magnetization form for the Navier-Stokes equation in $\mathbb{R}^n$ for the velocity field $u(x,t)$ of an incompressible fluid:

$$u_t + (u \cdot \nabla) u + \nabla p = \frac{1}{2}\Delta u, \qquad \operatorname{div} u = 0, \tag{7}$$

with initial condition $u|_{t=0} = u^0$. The variable $p(x,t)$ is the pressure and is necessary to impose the incompressibility condition $\operatorname{div} u = 0$.

For our purposes in this paper, it is convenient to rewrite (7) in new variables, the magnetization variables. These have been used to study the Euler equation by several authors, namely Buttke [4], Oseledets [16], Russo and Smereka [20], among others. We will follow Chorin [7] in the summary of results we present next.

The magnetization variable m is obtained by adding to the velocity field u a gradient

$$u = m + \nabla k.$$

The scalar function $k(x,t)$ is arbitrary at $t = 0$ and its evolution is chosen conveniently.

This transformation is a change of gauge, of which there are several possible choices, as discussed in [20]. Clearly, from m one can compute u by using the Leray projection on the divergence free vector fields:

$$u = \mathbb{P} m.$$

With an appropriate choice for k, the equation for the evolution of m is

$$\partial_t m_i + u_j D_j m_i + m_j D_i u_j = \frac{1}{2}\Delta m_i. \tag{8}$$

A main difference from (7) is that equation (8) does not involve pressure, nor $\operatorname{div} m = 0$. Furthermore, to any solution of (8) with $u = \mathbb{P}m$, corresponds a solution u to (7). In the other direction, to any solution of (7) and initial value of k there exists a solution of (8) such that $u = \mathbb{P}m$ for all times.

Acknowledgments

Supported in part by FCT/POCTI/FEDER and POCI/MAT/55745/2004.

References

1. V. Arnold, "Sur la géométrie différentielle des groupes de Lie de dimension infinie et ses applications à l'hydrodynamique des fluides parfaits", *Ann. Inst. Fourier (Grenoble)* **16**(fasc. 1), 319–361 (1966).
2. Barbara Busnello, Franco Flandoli,Marco Romito, "A probabilistic representation for the vorticity of a 3-dimensional viscous fluid and for general systems of parabolic equations".
3. Barbara Busnello, "A probabilistic approach to the two-dimensional Navier-Stokes equations", *Ann. Probab.* **27**(4), 1750–1780 (1999).
4. Tomas F. Buttke, "Velocity methods: Lagrangian numerical methods which preserve the Hamiltonian structure of incompressible fluid flow", in *Vortex flows and related numerical methods (Grenoble, 1992)*, volume 395 of *NATO Adv. Sci. Inst. Ser. C Math. Phys. Sci.*, Kluwer Acad. Publ., Dordrecht, 1993, pp. 39–57.
5. Ana Bela Cruzeiro, Fernanda Cipriano, preprint, 2005.
6. Alexandre Joel Chorin, "Numerical study of slightly viscous flow", *J. Fluid Mech.* **57**(4), 785–796 (1973).
7. Alexandre J. Chorin, *Vorticity and Turbulence*, volume 103 of *Applied Mathematical Sciences*. Springer-Verlag, New York, 1994.
8. P. Constantin, G. Iyer, "A stochastic Lagrangian representation of 3-dimensional incompressible Navier-Stokes equations" (2005).
9. Peter Constantin, "An Eulerian-Lagrangian approach for incompressible fluids: local theory", *J. Amer. Math. Soc.* **14**(2), 263–278 (electronic) (2001).
10. Peter Constantin, "An Eulerian-Lagrangian approach to the Navier-Stokes equations", *Comm. Math. Phys.* **216**(3), 663–686 (2001).
11. Peter Constantin, "Near identity transformations for the Navier-Stokes equations", In *Handbook of mathematical fluid dynamics* vol. II, North-Holland, Amsterdam, 2003, pp. 117–141.
12. David G. Ebin, Jerrold Marsden, "Groups of diffeomorphisms and the notion of an incompressible fluid", *Ann. of Math. (2)* **92**, 102–163 (1970).

13. Nassif Ghoussoub, "Anti-selfdual hamiltonians: Variational resolutions for navier-stokes and other nonlinear evolutions", preprint.
14. Diogo Aguiar Gomes, "A variational formulation for the Navier-Stokes equation" *Comm. Math. Phys.* **257**(1), 227–234 (2005).
15. Atsushi Inoue, Tadahisa Funaki, "On a new derivation of the Navier-Stokes equation", *Comm. Math. Phys.* **65**(1), 83–90 (1979).
16. V. I. Oseledets, "A new form of writing out the Navier-Stokes equation. Hamiltonian formalism" *Uspekhi Mat. Nauk* **44**(3(267)), 169–170 (1989).
17. Charles S. Peskin, "A random-walk interpretation of the incompressible Navier-Stokes equations", *Comm. Pure Appl. Math.* **38**(6), 845–852 (1985).
18. D. L. Rapoport, "Stochastic differential geometry and the random integration of the Navier-Stokes equations and the kinematic dynamo problem on smooth compact manifolds and Euclidean space", *Hadronic J.* **23**(6), 637–675 (2000).
19. D. L. Rapoport, "Random representations of viscous fluids and the passive magnetic fields transported on them", *Discrete Contin. Dynam. Systems* (added volume), 327–336 (2001). *Dynamical systems and differential equations (Kennesaw, GA, 2000).*
20. Giovanni Russo, Peter Smereka, "Impulse formulation of the Euler equations: general properties and numerical methods", *J. Fluid Mech.* **391**, 189–209 (1999).

DIOGO AGUIAR GOMES
Departamento de Matemática, Instituto Superior Técnico,
Av. Rovisco Pais, 1049-001 Lisboa, Portugal
dgomes@math.ist.utl.pt

CONVOLUTION CALCULUS ON WHITE NOISE SPACES AND FEYNMAN GRAPH REPRESENTATION OF GENERALIZED RENORMALIZATION FLOWS

HANNO GOTTSCHALK (*Universität Bonn*),
HABIB OUERDIANE (*Université de Tunis El Manar*),
BOUBAKER SMII (*Université de Tunis El Manar / Universität Bonn*)

In this note we outline some novel connections between the following fields: 1) convolution calculus on white noise spaces, 2) pseudo-differential operators and Lévy processes on infinite dimensional spaces, 3) Feynman graph representations of convolution semigroups, 4) generalized renormalization group flows and 5) the thermodynamic limit of particle systems.

Convolution semigroups on infinite dimensional spaces are the mathematical backbone of the Wilson-Polchinski formulation of renormalization group flows [6]. There, a "time" (in this context one should say: scale) dependent infinite dimensional heat equation

$$\frac{\partial}{\partial t} e^{-V_t^{\rm eff}(\phi)} = \dot{\Delta}_{t,T_0} e^{-V_t^{\rm eff}(\phi)}, \quad V_{T_0}^{\rm eff}(\phi) = V^{\rm initial}(\phi), \tag{1}$$

governs the flow of the effective action $V_t^{\rm eff}$ between two scales T, T_0, where T is the scale on which the system is observed and T_0 is the cut-off scale. The "time" dependent infinite dimensional Laplacian $\dot{\Delta}_{t,T_0} = \frac{\partial}{\partial_t}\Delta_{t,T_0}$ is defined by

$$\Delta_{t,T_0} = \int_{\mathbb{R}^d}\int_{\mathbb{R}^d} G_{t,T_0}(x,y)\frac{\delta}{\delta\phi(x)}\frac{\delta}{\delta\phi(y)}\,dx\,dy \tag{2}$$

with $G_{t,T_0} = G_t - G_{T_0}$ the covariance function of the Gaussian random field that has to be integrated out to intermediate between the random field with fundamental cut-off scale T_0 and covariance G_{T_0} and the random field with cut-off t and covariance G_t describing Gaussian fluctuations at some other scale $t \neq T_0$ ($t > T_0$ for ultra violet and $t < T_0$ for infra red problems).

The functional derivative $\frac{\delta}{\delta\phi(x)}$ acts on a function $F(\phi) = \sum_{n=0}^{\infty}\langle F_n, \phi^{\hat{\otimes} n}\rangle$ via

$$\frac{\delta}{\delta\phi(x)}F(\phi) = \sum_{n=0}^{\infty} n\langle D_x F_n, \phi^{\hat{\otimes} n-1}\rangle, \; D_x F_n(x_2,\ldots,x_n) = F_n(x,x_2,\ldots,x_n).$$

Here the kernel functions F_n are assumed to be symmetric.

The usefulness and mathematical beauty of the Wilson-Polchinski approach to renormalization is given by the following key features

1. The renormalization equation is the generating equation of a infinite dimensional diffusion process;
2. The perturbative solution of 1) can be represented as a sum over Feynman graphs.
3. Renormalization conditions at a scale T can be imposed by a change of the initial condition $V^{\rm initial}$ at scale T_0 leading to a finite theory if the cut-off T_0 is removed.

In this note we outline the generalization of (1) replacing the infinite dimensional Laplacian by a pseudo-differential operator in infinite dimensions. We show that under this generalization the key features of the renormalization group approach all prevail: Instead of infinite dimensional diffusion processes one obtains jump-diffusion type Lévy processes and instead of classical Feynman graphs generalized Feynman graphs studied in [2]. The idea of renormalization as given in item 3 above then carries over unchanged. We will illustrate this for the special case of particle systems in the continuum.

White noise analysis is a natural framework to rigorously formulate equations like (1) and their generalizations: In fact, following [3], we let $\mathcal{S} = \mathcal{S}(\mathbb{R}^d, \mathbb{C})$ be the space of rapidly decreasing test functions equipped with the Schwartz topology which is generated by an increasing sequence of Hilbert seminorms $\{|.|_p\}_{p\in\mathbb{N}}$ and let $\mathcal{S}'$ the dual of $\mathcal{S}$. By $\mathcal{S}_p$ we denote the closure of $\mathcal{S}$ w.r.t. $|.|_p$ and by $\mathcal{S}_{-p}$ the topological dual of $\mathcal{S}_p$. For $\theta(t)$, $t \leq 0$, a Young function (nonnegative, continuous, convex and strictly increasing s.t. $\lim_{t\to\infty} \theta(t)/t = \infty$) we set $\theta^*(t) = \sup_{x\geq 0}(xt - \theta(x))$, the Legendre transform of θ, which is another Young function.

Given a complex Banach space $(B, \|\,.\,\|)$, let $H(B)$ denote the space of entire function on B, i.e., the space of continuous functions from B to $\mathbb{C}$, whose restriction to all affine lines of B are entire on $\mathbb{C}$. Let $Exp(B, \theta, m)$ denote the space of all entire functions on B with exponential growth of order θ, and of finite type $m > 0$

$$Exp(B,\theta,m) = \left\{F \in H(B)\,;\, \|F\|_{\theta,m} = \sup_{z\in B} |F(z)| e^{-\theta(m\|z\|)} < +\infty\right\}. \quad (3)$$

In the following we consider the white noise test functions space

$$\mathcal{F}_\theta(\mathcal{S}') = \bigcap_{p\geq 0,\, m>0} Exp(\mathcal{S}_{-p}, \theta, m). \quad (4)$$

Let $\mathcal{F}_\theta(\mathcal{S}')^*$, the space of white noise distributions, be the strong dual of the space $\mathcal{F}_\theta(\mathcal{S}')$ equipped with the projective limit topology. For any $f \in \mathcal{S}$ and θ, the exponential $e^f : \mathcal{S}' \to \mathbb{C}$, $e^f(\phi) = e^{\langle \phi, f\rangle}$ is in $\mathcal{F}_\theta(\mathcal{S}')$. Thus, the Laplace transform $\mathcal{L} : \mathcal{F}_\theta(\mathcal{S}')^* \times \mathcal{S} \to \mathbb{R}$, $\mathcal{L}(\Phi)(f) = \langle \Phi, e^f\rangle$, is well defined. Recalling the definition of the space

$$\mathcal{G}_\theta(\mathcal{S}) = \bigcup_{p\geq 0,\, m>0} Exp(N_p, \theta, m) \tag{5}$$

which is equipped with the topology of the inductive limit, we get [3] that $\mathcal{L} : \mathcal{F}_\theta(\mathcal{S}')^* \to \mathcal{G}_{\theta^*}(\mathcal{S})$ is a topological isomorphism. Using the property of Young functions $\lim_{t\to\infty} \theta^*(t)/t = \infty$ it is easy to see that $\mathcal{G}_{\theta^*}$ is an algebra under multiplication. Thus, for $\Psi, \Phi \in \mathcal{F}_\theta(\mathcal{S}')^*$ one can define the convolution $\Psi * \Phi = \mathcal{L}^{-1}(\mathcal{L}(\Psi)\mathcal{L}(\Phi))$ as an element of $\mathcal{F}_\theta(\mathcal{S}')^*$.

Let us assume for a moment that $\lim_{t\to\infty} \theta(t)/t^2$ exists and is finite. Under this condition we have that $\mathcal{F}_\theta(\mathcal{S}') \to L^2_{\mathbb{C}}(\mathcal{S}'_{\mathbb{R}}, d\nu_0) \to \mathcal{F}_\theta(\mathcal{S}')^*$ is a Gelfand triplet, where ν_0 is the white noise measure, cf. [3]. Suppose that $\Psi \in \mathcal{F}_\theta(\mathcal{S}')$ has a Taylor series $\Psi(\phi) = \sum_{n=0}^N \langle \Psi_n, \phi^{\hat{\otimes} n}\rangle$ with $\Psi_n \in \mathcal{S}^{\hat{\otimes} n}$. One can then show that $\Psi * \Phi \in \mathcal{F}_\theta(\mathcal{S}')$ and $\Psi * \Phi(\phi) = \Psi(\frac{\delta}{\delta\phi})\Phi(\phi)$ with

$$\Psi\left(\frac{\delta}{\delta\phi}\right) = \sum_{n=0}^{\infty} \int_{\mathbb{R}^{dn}} \Psi_n(x_1, \ldots, x_n) \frac{\delta}{\delta\phi(x_1)} \cdots \frac{\delta}{\delta\phi(x_n)}\, dx_1 \cdots dx_n\,. \tag{6}$$

Here the rule for the evaluation of the pseudo-differential operator $\Psi(\frac{\delta}{\delta\phi})$ is that first the n-th order differential operators are applied to Φ and then the result is summed up over all n. Hence we see that the equation

$$\frac{\partial}{\partial t} e^{-V_t^{\text{eff}}} = \dot{\Psi}_{t,T_0}\left(\frac{\delta}{\delta\phi}\right) e^{-V_t^{\text{eff}}} = \dot{\Psi}_{t,T_0} * e^{-V_t^{\text{eff}}}, \quad V_{T_0}^{\text{eff}} = V^{\text{initial}}, \tag{7}$$

is the correct generalization of the renormalization flow equation (1). If now $\dot{\Psi}_{t,T_0} : \mathbb{R}_+ \to \mathcal{F}_\theta(\mathcal{S}')^*$ is continuous and $e^{-V^{\text{initial}}} \in \mathcal{F}_\theta(\mathcal{S}')^*$, it has been proven in [5] that (7) has an unique solution in $\mathcal{F}_{(e^{\theta^*}-1)^*}(\mathcal{S}')^*$, namely

$$e^{-V_t^{\text{eff}}} = \mathcal{L}^{-1}\left(e^{\int_{T_0}^t \mathcal{L}(\dot{\Psi}_{s,T_0})\, ds}\right) * e^{-V^{\text{initial}}}. \tag{8}$$

Again, the above solution is of particular interest if a probabilistic interpretation can be given. This is the case when $\mathcal{L}(\dot{\Psi}_{t,T_0})$ for every t is a conditionally positive function, i.e. $\mathcal{L}(\Psi_{t,T_0})(0) = 0$,

$$\sum_{j,l=1}^n \mathcal{L}(\dot{\Psi}_{t,T_0})(f_j + f_l) z_j \bar{z}_l \leq 0 \quad \forall t < T_0, n \in \mathbb{N}, f_l \in \mathcal{S}_{\mathbb{R}}, z_l \in \mathbb{C} : \sum_{l=1}^n z_l = 0 \tag{9}$$

and the opposite inequality holds for $t > T_0$. Under these conditions, by the Bochner-Minlos theorem, the transition kernel $\mathcal{L}^{-1}\big(e^{\int_{T_0}^t \mathcal{L}(\dot{\Psi}_{s,T_0})\,ds}\big)$ is a family of probability measures on $\mathcal{S}'$ that fulfills the Chapman-Kolmogorov equations and thus defines a stochastic process with state space $\mathcal{S}'$. In general, this process will be of jump-diffusion type, as it follows from the Lévy-Itô decomposition of conditionally positive definite functions.

Let us now come to the Feynman graph expansion. We take an initial condition of the type

$$V^{\text{initial}}(\phi) = \sum_{p=0}^{\bar{p}} \langle \lambda^{(p)}, \phi^{\otimes p} \rangle \tag{10}$$

with kernel (vertex) functions $\lambda^{(p)} \in \mathcal{S}^{\hat{\otimes} p}$, for simplicity (this condition can clearly be relaxed). It is also assumed that $V^{\text{initial}}(\phi) \geq -C$ for some $C > 0$ and all $\phi \in \mathcal{S}'$. Then, $e^{-V^{\text{initial}}} \in L^2(\mathcal{S}', d\nu_0)$. If now $\theta(t)$ fulfills $\lim_{t\to\infty} \theta(t)/t^2 = c < \infty$, then by the theorem cited above, the solution of (7) exists.

The next step is to expand (7) in a formal power series in V^{initial}. At least in the case where, for t fixed, $\mathcal{L}^{-1}\big(e^{\int_{T_0}^t \mathcal{L}(\dot{\Psi}_{s,T_0})\,ds}\big)$ is a measure on $\mathcal{S}'_{\mathbb{R}}$, this expansion is an asymptotic series, cf. [2, Lemma 2.2]. We note that the Laplace transform of a white noise distribution is an analytic function [3]. One can thus consider the Taylor series in f of $e^{\int_{T_0}^t \mathcal{L}(\dot{\Psi}_{s,T_0})(f)\,ds}$ at zero given by

$$\sum_{m\geq 0} \frac{1}{m!} \langle m_{n,t,T_0}, f^{\hat{\otimes} n} \rangle \tag{11}$$

for $f \in \mathcal{S}$ with $m_{n,t,T_0} \in (\mathcal{S}')^{\hat{\otimes} n}$ the n-th moment. The connected moment functions, m^c_{n,t,T_0} by definition are the Taylor coefficients of the logarithm of the generating functional of the moment functions, i.e.,

$$\int_{T_0}^t \mathcal{L}(\dot{\Psi}_{s,T_0})(f)\,ds = \sum_{m\geq 0} \frac{1}{m!} \langle m^c_{n,t,T_0}, f^{\hat{\otimes} n} \rangle \,. \tag{12}$$

The well-known linked cluster theorem, cf. e.g. [2, Appendix A], then gives the combinatorial relation between moments and connected moments, namely

$$m_{n,t,T_0}(x_1, \ldots, x_n) = \sum_{\substack{I \in \mathcal{P}(1,\ldots,n) \\ I = \{I_1,\ldots,I_k\}}} \prod_{l=1}^{k} m^c_{n_l,t,T_o}(x_{j_l^1}, \ldots, x_{j_l^{n_l}}) \,, \tag{13}$$

where $\mathcal{P}(1,\ldots,n)$ is the set of all partitions of $\{1,\ldots,n\}$ into disjoint non empty subsets $I_1,\ldots,I_k$, $k \in \mathbb{N}$ arbitrary, $I_l = \{j_l^1,\ldots,j_l^{n_l}\}$. After these preparations one obtains by straight forward calculation

$$\begin{aligned}
&e^{-V_t^{\text{eff}}(\phi)} \\
&= \sum_{m=0}^{\infty} \frac{(-1)^m}{m!} \mathcal{L}^{-1}\left(e^{\int_{T_0}^t \mathcal{L}(\dot{\Psi}_{s,T_0})\,ds}\right) * (V^{\text{initial}})^m(\phi) \\
&= \sum_{m=0}^{\infty} \frac{(-1)^m}{m!} \sum_{p_1,\ldots,p_m=0}^{\bar{p}} \int_{\mathbb{R}^{d\#\Omega}} \lambda^{(p_1)}(x_1^1 \cdots x_{p_1}^1) \cdots \lambda^{(p_m)}(x_1^m, \cdots, x_{p_m}^m) \\
&\quad \times \sum_{K\subset\Omega} \prod_{(s,q)\in K} (-\phi(x_s^q))\, m_{t,T_0}(\Omega\setminus K) \otimes_{(s,q)\in\Omega} dx_s^q \\
&= \sum_{m=0}^{\infty} \frac{(-1)^m}{m!} \sum_{p_1,\ldots,p_m=0}^{\bar{p}} \int_{\mathbb{R}^{d\#\Omega}} \lambda^{(p_1)}(x_1^1 \cdots x_{p_1}^1) \cdots \lambda^{(p_m)}(x_1^m, \cdots, x_{p_m}^m) \\
&\quad \times \sum_{K\subset\Omega} \prod_{(s,q)\in K} (-\phi(x_s^q)) \sum_{\substack{I\in\mathcal{P}(\Omega\setminus K)\\ I=\{I_1,\ldots,I_k\}}} \prod_{l=1}^{k} m_{t,T_0}^c(I_l) \otimes_{(s,q)\in\Omega} dx_s^q.
\end{aligned} \tag{14}$$

Here we used the following notation: $\Omega = \Omega(p_1,\ldots,p_m) = \{(1,1),\ldots,(1,p_1),\ldots,(m,1),\ldots,(m,p_m)\}$ and for $A \subseteq \Omega$, $A = \{(s_1,q_1),\ldots,(s_r,q_r)\}$, $m_{t,T_0}^{(c)}(A) = m_{r,t,T_0}^{(c)}(x_{s_1}^{q_1},\ldots,x_{s_r}^{q_r})$. $\mathcal{P}(\Omega\setminus K)$ is the set of partitions of $\Omega\setminus K$. We also made use of the fact that $(V^{\text{initial}})^m(\phi)$, being a polynomial with test functions as coefficients, is a white noise test function and that the convolution between a white noise distribution Φ and a white noise test function Θ is $\Phi * \Theta(\phi) = \Phi(\Theta_\phi)$ where $\Theta_\phi(\varphi) = \Theta(\varphi - \phi)$ is a shift.

Generalized Feynman graphs can now be used to order the combinatorial sum on the r.h.s. of (14). A generalized (amputated) Feynman graph is a graph with three types of vertices, called inner full $\bullet$, inner empty $\circ$ and outer empty $\otimes$ vertices, respectively. By definition full vertices are distinguishable and have distinguishable legs whereas empty vertices are non distinguishable and have non distinguishable legs. Outer empty vertices are met by one edge only. Edges are non directed and connect full and empty (inner and outer) vertices, but never connect two full or two empty vertices. The set of generalized Feynman graphs with m inner full vertices with $p_1,\ldots,p_m$ the number of legs of the inner full vertices such that $p_j \leq \bar{p}$ and $\lambda^{(p_j)} \neq 0$, $j = 1,\ldots,m$, is denoted by $\bar{F}(m)$.

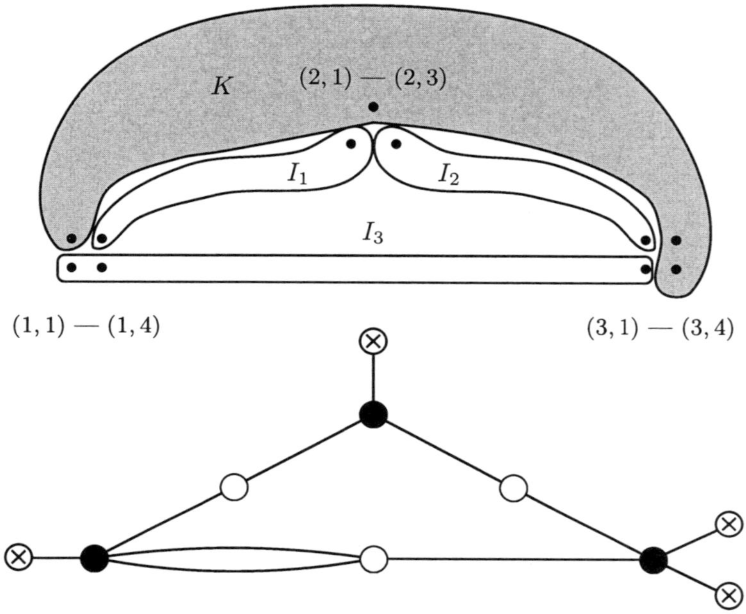

Figure 1. Construction of a generalized Feynman graph from the set K and the partition $I = \{I_1, I_2, I_3\}$

To obtain the connection with (14) we consider an example where $m = 3$, $p_1 = 4, p_2 = 3$ and $p_3 = 4$. For each element in $\Omega = \Omega(4, 3, 4)$ we draw a point s.t. points belonging to the same p_l are drawn close together. Then we choose a subset K and a partition I, cf. figure 1 (top). The generalized Feynman graph can now be obtained by representing each collection of points by a full inner vertex with p_l legs, for each set I_l in the partition we draw a inner empty vertex connected to the legs of the inner full vertices corresponding to the points in I_l. Finally we draw an outer empty vertex connected to the leg of an inner full vertex corresponding to each point in K. We then obtain the generalized Feynman graph figure 1 (bottom). In this way, for fixed m, one obtains a one to one correspondence between the index set of the sum in (14) and $\bar{F}(m)$.

We want to calculate the contribution to (14) directly from the graph $G \in \bar{F}(m)$ without the detour through the above one to one correspondence. This is accomplished by the following Feynman rules: Attribute a vector

$x \in \mathbb{R}^d$ to each leg of a inner full vertex. For each inner full vertex with p legs multiply with $\lambda^{(p)}$ evaluated at the vectors attributed to the legs of that vertex. For any inner empty vertex with l legs, multiply with a connected moment function m^c_{l,t,T_0} evaluated with the l arguments corresponding to the legs that this inner empty vertex is connected with. For each outer empty vertex multiply with $-\phi(x)$ where x is the argument of the leg of the inner full vertex that the outer empty vertex is connected with. Finally integrate $\int_{\mathbb{R}^d} \cdots dx$ over all the arguments x that have been used to label the legs of the inner full vertices. In this way one obtains the analytic value $\mathcal{V}[G](t, T_0, \phi)$. The perturbative solution of (7) then takes the form

$$e^{-V_t^{\rm eff}(\phi)} = \sum_{m=0}^{\infty} \frac{(-1)^m}{m!} \sum_{G \in \bar{F}(m)} \mathcal{V}[G](t, T_0, \phi), \tag{15}$$

where the identity is in the sense of formal power series. The linked cluster theorem for generalized Feynman graphs proven in [2, 4] then implies that $V_t^{\rm eff}$ can be calculated as a sum over connected Feynman graphs

$$-V_t^{\rm eff}(\phi) = \sum_{m=0}^{\infty} \frac{(-1)^m}{m!} \sum_{G \in \bar{F}_c(m)} \mathcal{V}[G](t, T_0, \phi). \tag{16}$$

Let us now apply the above renormalization group scheme to the problem of taking the thermodynamic limit of a particle system. To this aim let $T = 0$ and

$$\mathcal{L}(\mu_t)(f) = e^{\int_{\mathbb{R}^d} \int_{[-c,c]} \left(e^{sf(x)}-1\right) dr(s) \sigma_t(x) dx} \tag{17}$$

where r is a probability measure on $[-c, c]$, $0 < c < \infty$, $\sigma_t(x) = \sigma(x/t)$ where σ is a continuously differentiable function with support in the unit ball and $\nabla\sigma(0) = 0$ such that $\sigma(0) = z > 0$. $\mathcal{L}(\mu_t)$ at the same time is the Laplace transform of the Poisson measure μ_t representing a system of non-interacting, charged particles in the grand canonical ensemble with intensity measure (local density) σ_t, see e.g. [1]. and a white noise distribution $\mu_t \in \mathcal{F}_\theta(\mathcal{S}')^*$ for any θ (due to the linear exponential growth of (17) in f we have that the r.h.s. is in $\mathcal{G}_{\theta^*}(\mathcal{S})$ for any θ), cf. [3]. Both objects can thus be identified. Furthermore we assume that $\sigma(\alpha x)$ is monotonically decreasing in α for $\alpha > 0$. Then $\dot{\sigma}_t(x) = d\sigma_t(x)/dt \geq 0$ for $t > 0, x \in \mathbb{R}^d$. It is then standard to show that

$$\mathcal{L}(\dot{\Psi}_t)(f) = -\int_{\mathbb{R}^d} \int_{[-c,c]} \left(e^{sf(x)} - 1\right) dr(s)\, \dot{\sigma}_t(x)\, dx \tag{18}$$

in fact fulfills (9) for all $t < T_0$. Thus the pseudo differential operator $\dot{\Psi}(\delta/\delta\phi)$ is the generator of a jump-diffusion process with state space $\mathcal{S}'$ with backward time direction. Let $\lambda^{(p)} = \lambda^{(p)}(x_1, \ldots, x_p)$ be a set of C^∞ functions that are of rapid decrease in the difference variables $x_i - x_j$, $i \neq j$. For a distribution ϕ of compact support we define $V^{\text{initial}}(\phi)$ as in (10). We assume that $V^{\text{initial}}(\phi) > -C$ for each such ϕ.

Let $\phi \in \mathcal{S}'$ have compact support. The non normalized correlation functional $\rho_t(\phi)$ of the particle system with infra-red cut-off t is defined as

$$\rho_t(\phi) = \int_{\mathcal{S}'} e^{-V^{\text{initial}}(\varphi+\phi)}\, d\mu_t(\varphi) = \mu_t * e^{-V^{\text{initial}}}(\phi). \tag{19}$$

Note that μ_{T_0} has support on the distributions supported on a ball of diameter T_0, B_{T_0}. Therefore, for ϕ fixed and $t < T_0$ one can replace $\lambda^p(x_1, \ldots, x_n)$ with $\lambda^{(p)}(x_1, \ldots, x_n) \prod_{l=1}^n \chi(x_l)$ with χ being a test function that is constantly one on $\operatorname{supp}\phi \cup B_{T_0}$ without changing (19). Under this replacement, V^{initial} meets the conditions from above that $\lambda^{(p)} \in \mathcal{S}^{\hat{\otimes} p}$. Furthermore, under this replacement in V^{initial}, $e^{-V^{\text{initial}}} \in L^2(\mathcal{S}', d\nu_0)$ and hence $e^{-V^{\text{initial}}} \in \mathcal{F}_\theta(\mathcal{S}')^*$ if $\lim_{t\to\infty} \theta(t)/t^2$ finite. Furthermore, one can argue as above to see that $\dot{\Psi}_t$ is in $\mathcal{F}_\theta(\mathcal{S}')^*$ for θ arbitrary. Thereby, the requirements of white noise convolution calculus are met. We shall neglect the inessential distinction between V^{initial} and its replaced version in the following.

From (17)–(19) we see that the non normalized correlation functional ρ_t fulfills the renormalization group equation (7). The thermodynamic limit, which is achieved as $T_0 \to \infty$, is thus governed by this equation (and thus by a Lévy process with infinitely dimensional state space).

Let us now come to the issue of the normalization of ρ_T at a time $T = 0$, for simplicity. A normalized correlation functional should fulfill $\rho_T(0) = 1$. But $-V_T^{\text{eff}}(0) = \log \rho_T(0) \sim T_0^d$ in our case where the divergent (as $T_0 \to \infty$) parts originate from the so-called vacuum to vacuum diagrams, i.e., such Feynman graphs in $\bar{F}_c(m)$ that do not have outer empty vertices [2, Thm. 6.6]. All other contributions remain finite in the limit $T_0 \to \infty$ [2, Section 7]. The normalization $-V_T^{\text{eff}}(0) = 0$ can now be achieved perturbatively by re-defining V^{initial} by a counter term

$$\delta\lambda_{T_0}^{(0)} = \sum_{m \geq 0} \frac{(-1)^m}{m!} \sum_{G \in \bar{F}_c^{\text{vac. to vac.}}(m)} \mathcal{V}[G](0, T_0, 0) \tag{20}$$

and replacing $\lambda^{(0)}$ by $\lambda^{(0)} - \delta\lambda_T^{(0)}$ since this removes properly the vacuum to vacuum diagrams at any order m of perturbation theory of V_t^{eff} and other diagrams give vanishing contribution for $\phi = 0$.

The (perturbative) thermodynamic limit $T_0 \to \infty$ of $\rho_T^{\text{ren}}(\phi)$ can now be taken achieving at once the finiteness of the perturbation expansion and the normalization of $\rho^{\text{ren}}(\phi)$ since $\mathcal{V}[G](T = 0, T_0, \phi)$ converges as $T_0 \to \infty$ for $G \in \bar{F}_c(m)$ not a vacuum to vacuum diagram, cf. [2, Section 7].

Of course, the above renormalization problem is rather trivial as the particle system only has short range p-body forces for $p \leq \bar{p}$. But having put the problem of TD limits of particles in the continuum in the language of the (generalized) renormalization group now paves the way to the use of typical renormalization techniques, as e.g. differential inequalities and inductive proofs of renormalizability order by order in perturbation theory [6], to tackle less trivial problems in the thermodynamics of particle systems with long range forces.

Acknowledgements

H. Gottschalk has been supported through the German Research Council (DFG) project "Stochastic methods in QFT", while B. Smii has been supported by the German Academic Exchange Service (DAAD); the authors gratefully acknowledge this. H. O. would like to thank S. Albeverio for for his warm hospitality and DAAD and DFG financial support when being in Bonn. The authors also thank Sergio Albeverio for his encouragement and interesting discussions.

References

1. S. Albeverio, H. Gottschalk, M. W. Yoshida, "Systems of classical particles in the grand canonical ensemble, scaling limits and quantum field theory", *Rev. Math. Phys.* **17**, no. 2, 356–369 (2005), `arXiv:math-ph/0601021`.
2. S. H. Djah, H. Gottschalk, H. Ouerdiane, "Feynman graph representation of the perturbation series for general functional measures", *J. Funct. Anal.* **227**, 153–187 (2005), `arXiv:math-ph/0408031`.
3. R. Gannoun, R. Hachaichi, H. Ouerdiane, A. Rezgui, "Un theéorème de dualité entre espaces de fonctions holomorphes à croissance exponentielle", *Journ. Funct. Anal.* **171**, 1–14 (2000).
4. H. Gottschalk, B. Smii, H. Thaler, "The Feynman graph representation of convolution semigroups and applications to Lévy statistics", Bonn preprint 2005, `arXiv:math.PR/0601278`.
5. H. Ouerdiane, N. Privault, "Asymptotic estimates for white noise distributions", *C. R. Acad. Sci. Paris, Ser. I*, **338**, 799–804 (2004).
6. M. Salmhofer, *Renormalization*, Springer Verlag, Berlin/Heidelberg, 1999.

Hanno Gottschalk
Institut für angewandte Mathematik,
Wegelerstr. 6, D-51373 Bonn, Germany
gottscha@wiener.iam.uni-bonn.de

Habib Ouerdiane
Département de Mathématique, Université de Tunis El Manar,
Campus Universitaire, TN-1006, Tunis
habib.ouerdiane@fst.rnu.tn

Boubaker Smii
Institut für angewandte Mathematik,
Wegelerstr. 6, D-51373 Bonn, Germany
boubaker@wiener.iam.uni-bonn.de

CHARACTERIZATIONS OF STANDARD NOISES AND APPLICATIONS

Takeyuki Hida (*Meijo University, Nagoya*),
Si Si (*Aichi Prefectural University, Aichi-ken*)

1. Introduction

As a background of white noise analysis, we take two basic standard noises; one is Gaussian noise (white noise) and the other is Poisson noise. The second is immediately generalized to compound Poisson noise.

Those noises are idealized elemental stochastic processes. The reason why we take those noises as basic object comes from our idea of stochastic analysis.

Suppose we are given a complex random system to be analyzed, we shall proceed as follows. First form an elemental system of random variables that contain the same information as the given random system, i.e., Reduction. We are therefore given a system of independent random variables which can be taken to be the variables and the given random system is to be represented as a functional of those variables, i.e., Synthesis. Finally follows Analysis of the function in order to investigate the random system in question.

With this plan, we first discuss elemental random systems, to fix the idea we take elemental noises. The basic noises are Gaussian or Poisson, as we can understand by the Lévy decomposition (or Lévy-Itô decomposition) of a Lévy process. It leads us to find out characteristic properties of them for detailed investigation of the random system expressed as functions of noises.

2. Invariance of the noises

Following the idea of reduction, we take two standard systems of idealized elemental random variables. Such a system should consist of i.i.d. (independent identically distributed) and be parameterized by the time variable t. As is suggested by the book by Gel'fand-Vilenkin [5], it is natural to consider a (Gaussian) white noise and Poisson noise as elemental noises. They are stationary, generalized stochastic process. They have flat spectrum, so that they are called white noise. Customary, Gaussian white noise is called simply *white noise* and Poisson type elemental system is called *Poisson noise.*

2.1. Two noises

The two noises may intuitively be defined as time derivatives of a Brownian motion $B(t)$ and a Poisson process $P(t)$, respectively. Sample functions of their time derivatives are no more ordinary functions, but generalized functions. Their probability distributions are therefore introduced in the space of generalized functions, denoted by E^*, which is a dual space of some nuclear space E $(\subset L^2(R^1))$.

Given $\dot{B}(t)$ and $\dot{P}(t)$, their characteristic functionals are easily computed as follows:

$$C^G(\xi) = \exp\left[-\frac{1}{2}\|\xi\|^2\right], \quad \xi \in E,$$

and

$$C^P(\xi) = \exp\left[\lambda \int (e^{i\xi(t)} - 1)\, dt\right].$$

2.2. (Gaussian) white noise

We start with the characteristic functional $C^G(\xi)$. So there is a probability measure μ on a measurable space $(E^*, \mathbf{B})$, where E^* is a space of generalized functions and where $\mathbf{B}$ is a sigma-field generated by cylinder subsets of E^*, in such a way that

$$C^G(\xi) = \int_{E^*} \exp[i\langle x, \xi\rangle]\, d\mu(x). \tag{1}$$

To describe *invariance* of μ, we must introduce the *infinite dimensional rotation group.*

Take the nuclear space E as before and define a *rotation* g of E to be

i) a linear isomorphism of E and
ii) orthogonal : $\|g\xi\| = \|\xi\|$,

where $\|\cdot\|$ is the $L^2(R^1)$-norm.

The collection of rotations forms a group which is called the *infinite dimensional rotation group* and is denoted by $O(E)$. We introduce the compact-open topology to $O(E)$ to let it be a topological group.

The collection of adjoint transformations g^* of $g \in O(E)$ forms a group denoted by $O^*(E^*)$. The correspondence

$$g \to (g^*)^{-1}$$

is bijective, so that we can topologize $O^*(E^*)$ to be a topological group isomorphic to $O(E)$.

The transformation g^* is $\mathbf{B}$-measurable, so that the product $g^*\mu$ is defined. The following assertions (Proposition 2.1 and Proposition 2.2) are known, however we list to remind them.

Proposition 2.1. *The white noise measure μ is $O^*(E^*)$-invariant. Namely, for any g^* in $O^*(E^*)$ it holds that*

$$g^*\mu = \mu.$$

This is the reason why the infinite dimensional rotation group plays an important role in white noise analysis. In fact, by the use of the rotation group we can carry on an infinite dimensional harmonic analysis. This is one of the big advantages of our analysis.

A converse direction of Proposition 2.1 is significant. To claim the assertion, we provide a notation. The probability measure determined by the characteristic functional $\exp[-\frac{\sigma^2}{2}\|\xi\|^2]$ is denoted by μ_σ and is called a Gaussian measure with variance σ^2.

Proposition 2.2. *The probability measure ν determined by a characteristic functional $C(\xi)$ which is $O(E)$-invariant, i.e., $C(g\xi) = C(\xi)$ for every g in $O(E)$, then ν is expressed in the form*

$$\nu = a\delta_0 + \int_{(0,\infty)} \mu_\sigma \, dm(\sigma), \tag{2}$$

where $a \geq 0$ and dm is a bounded Borel measure on $(0, \infty)$.

If we further assume that ν is non-trivial and *atomic* (or $O^*(E^*)$-ergodic), we are given a single Gaussian measure μ_σ, that is to be an *elemental* noise.

Our infinite dimensional rotation group $O(E)$ involves two important families of subgroups. To define the first family, we use a complete orthonormal system, say $\{\xi_n\}$. We require that each ξ_n is a member of E, otherwise quite arbitrary. With the help of the ξ_n, we can define, in an obvious manner, a subgroup G_n isomorphic to $SO(n)$. Also, their inductive limit G_∞ can be defined. It is noted that this family contains much bigger subgroups like the Lévy group which involves essentially infinite dimensional rotations (see e.g. [10, chapter 5]).

Another family comes from a group of diffeomorphisms acting on the parameter space R^1. This class consists of one-parameter subgroups, called *whiskers* which describe profound probabilistic properties by using geometric structure of the parameter space. There is, among others, an interesting subgroup, denoted by $C(1)$, which is generated by whiskers and is isomorphic to the conformal group acting on R^1. It illustrates latent traits of white noise; e.g. it shows the *projective invariance* of Brownian motion. In the higher dimensional parameter case, say R^d, the family is much larger, in particular, we can define the conformal group $C(d)$ which is locally isomorphic to $SO(d+1,1)$. (For detail see [10].)

With this review, given as a background, we come to a generalization.

2.3. Two-dimensional valued white noise.

Having been suggested by S. Albeverio and A. Sengupta [3], we discuss two-dimensional valued white noise and applications to path integral.

Assume that the collection of (x,y)'s is the product space $E_x^* \times E_y^*$ of the spaces E_x^* and E_y^* consisting of generalized functions, respectively. The product measure $\nu(dz) = \mu_X(dx) \times \mu_Y(dy)$, with $dz = dx \wedge dy$ is introduced to $E_x^* \times E_y^*$.

It is obvious that

Proposition 2.3. *The probability measure $\nu(dxdy)$ is invariant under the rotations acting on (x,y)-space.*

In particular, under the transformation

$$z = (x,y) \ \rightarrow \ (x+y, x-y)/\sqrt{2} = (x', y')$$

the measure $\nu(dz)$ is kept invariant.

We may understand that the renormalized inner product $\langle x, y\rangle$, $x \in E_x^*, y \in E_y^*$ is defined below:

$$\langle x, y\rangle = \frac{1}{2}\left(\|x+y\|^2 - \|x-y\|^2\right),$$

where $\|x\| = \sqrt{\int : x(t)^2 : dt}$. We can see that $(x+y)$ and $(x-y)$ are independent under the Gaussian measure ν. Hence, there is no difficulty to deal with generalized functionals of $\langle x, y\rangle$. As an example, the functional $e^{ic\langle x,y\rangle}$ can be reduced to the product of independent Gauss kernels of the form

$$\varphi_1(x') = N \exp\left[c \int : x'(t)^2 : dt\right],$$

and the result is of the form $\varphi_1(x')\varphi_2(y')$.

Thus, we have the following theorem. Although the assertion seems obvious, some background is necessary, as we have seen above.

Theorem 2.1. *The integral*

$$\int e^{i\langle x,y\rangle}\Psi(z)\,dv(z)$$

is well defined if Ψ *is a test functional of* z.

There are various publications in this direction by many authors; for instance, S. Albeverio, A. Hahn and A. Sengupta [2] and others. It is our hope that interesting developments will be obtained in line with white noise analysis.

It is easy to generalize the results to higher dimensional valued white noise. Take a tensor product $E_1 \otimes \cdots \otimes E_n$ of spaces E_j, each of which is a copy of a nuclear space E. A product Gaussian measure ν is introduced on the space $E_1^* \otimes \cdots \otimes E_n^*$. We define the direct product $O^*(E_1^*) \times \cdots \times O^*(E_n^*)$ under which ν is invariant.

Theorem 2.2. *The direct product group* $O^*(E_1^*) \times \cdots \times O^*(E_n^*)$ *and the rotation group on* $E_1^* \otimes \cdots \otimes E_n^*$ *give invariances of the measure* ν.

2.4. Gauge transformations

The infinite dimensional rotation group $O(E)$ can be complexified to have the infinite dimensional unitary group $U(E_c)$, where E_c is the complexification of the nuclear space E. Many interesting subgroups have been discovered so far (see [7], [10]). This note extends the Heisenberg group and shows an invariance of complex white noise.

The Heisenberg group that we have introduced in [7] involves three one-parameter groups: Let ζ be in E_c.

a. The gauge transformation I_t is defined by

$$I_t : \zeta(u) \to (I_t\zeta)(u) = e^{it}\zeta(u).$$

b. The shift S_t is

$$S_t : \zeta(u) \to (S_t\zeta)(u) = \zeta(u-t).$$

c. Multiplication π_t is such a transformation that

$$\pi_t : \zeta(u) \to (\pi_t\zeta)(u) = e^{iut}\zeta(u).$$

We now define $I_{\alpha t}$ by

$$I_{\alpha t} : \zeta(u) \to (I_{\alpha t}\zeta)(u) = e^{i\alpha(u)t}\zeta(u),$$

where α is a member of the (real) Schwartz space S.

d. The collection $\{I_{\alpha t}, \alpha \in S\}$ forms a subgroup of $U(E_c)$ and is called *S-gauge transformation* group. (It is *abelian* in Physics.)

We now list the infinitesimal generators of the whiskers in a, b, c, and d.

$$iI, \quad s = -\frac{d}{du}, \quad \pi = iu\cdot, \quad i\alpha \in S.$$

The following theorem can be proved without any difficulty.

Theorem 2.3.

i) The generators listed above form a Lie algebra.
ii) The S-gauge transformations act on $z = (x, y)$-space as rotations, and ν measure is invariant under those rotations.

The assertion ii) is a generalization of what we have discussed in section 2.3.

3. Poisson noise

We shall discuss characteristics of Poisson noise, although they are latent, by comparing with the characteristics of Gaussian noise (white noise). The Gaussian case has been discussed in [8]. The Poisson case can be discussed in rather similar manner, but somewhat different to be interesting.

A general setup is this. We follow [15].

The characteristic functional of Poisson noise is

$$C^P(\xi) = \exp\left[\lambda \int_R (e^{i\xi(t)} - 1)dt\right], \quad \xi \in E.$$

With this in mind we wish to discover latent characteristics of Poisson noise.

Let a Poisson process $P(t, \omega)$, $\omega \in \Omega(P)$, be given. We restrict our attention to the unit time interval $[0, 1]$. Setting

$$A_n = \{\omega, P(1, \omega) = n\},$$

we are given a measurable partition $\{A_n, n \geq 0, \}$ of $\Omega(P)$

$$\cup A_n = \Omega, \quad A_n \cap A_m = \emptyset, \quad n \neq m \text{ a.s.}$$

We have

$$P(A_n) = \frac{\lambda^n}{n!} e^{-\lambda}, \quad n \geq 0. \tag{3}$$

Let $\tau_j(\omega)$ be the j-th jump point of $P(t, \omega), t \in [0, 1]$. If ω is restricted to A_n, there are exactly n jump points, $\tau_1, \ldots, \tau_n$. Those jump points are distributed independently on the unit time interval. The joint probability density $p(x_1, \ldots, x_n)$ is invariant under the permutation of the variables x_j. Obvious assertion is given in other words.

Proposition 3.1. *Let P_n be the conditional probability measure on A_n. The conditional joint probability density $p(x_1, \ldots, x_n)$ of jump points is invariant under the symmetric group $S(n)$ that acts as permutation group of variables.*

We can further show another characteristic of Poisson noise by restricting the sample space to be A_n. If we may consider n discrete random points that are distributed over the unit time interval, in general, to form a point process, the case of Poisson noise is the optimal case where the system of those n points enjoys maximum entropy. In this sense Poisson noise has a latent optimality in the sense of information theory.

The *converse* direction of the study is as follows. Indeed, this is more interesting part.

Start with a partition $\{A_n, n \geq 0\}$ of Ω. The probability $P(A_n)$, which is non zero, will be determined later.

Introduce random variables on A_n as many as n, denoted by $\tau_1, \ldots, \tau_n$. Let them be distributed over the unit time interval. Now consider the case where

1) the joint probability density $q(x_1, \ldots, x_n)$ of the random vector $(\tau_1, \ldots, \tau_n)$ is invariant under the symmetric group $S(n)$ acting as permutation of the component of the vector, and
2) the random vector has maximum entropy.

Proposition 3.2. *Under the assumptions 1) and 2), the joint distribution q is in agreement with the p introduced above.*

We are now given a characterization of Poisson noise provided that the probability $P(A_n)$ is determined as in (3). There are many reasonable ways to determine these probabilities like in [15], however we do not go into details.

Finally we come to the conclusion to observe that the conditional characteristic functional $C_n^P(\xi)$ is expressed in the form

$$C_n^P(\xi) = \left(\int_0^1 e^{i\xi(t)}\, dt \right)^n,$$

and the weighted sum, by using the values of the $P(A_n)$'s, is

$$\sum_{n=0}^{\infty} e^{-\lambda} \frac{\lambda^n}{n!} C_n^P(\xi) = \exp\left[\lambda \int_0^1 (e^{i\xi(t)} - 1) dt \right]. \tag{4}$$

The sum (4) is the characteristic functional of Poisson noise with parameter space $[0, 1]$.

In this way, a characterization, using latent properties, of Poisson noise has been given.

4. Application to information sociology

There are various applications of the characterization of Gaussian and Poisson noises discussed in this paper. Among others, we consider an application to information sociology, where stable distribution appears.

Then, we may think that the data may be a distribution of a random variable of a stable process evaluated at a fixed instant. Namely, one should check that the random phenomena in question can be embedded in a stable stochastic process of some exponent α. Suppose this is confirmed in some way. It is a superposition of Poisson processes with various magnitudes of jump. Anyhow elemental process is a Poisson process. According to our discussion, one can speak of latent characteristics such as optimality and symmetry. It is interesting to find that those characteristic properties correspond to the properties of the actual phenomena

Suppose we are given observed data, the cumulative distribution function will be close to a stable distribution.

At the same time one can think of the weight to have a superposition of Poisson noises, i.e., the Lévy measure, in order to find some relation to the structure of random phenomena. The Lévy measure for a stable process with exponent α is proportional to $|u|^{-(\alpha+1)}$ which is the weight associated to the Poisson noise with magnitude u of jump. This may tell us some meaning of the phenomena.

References

1. L. Accardi et al., eds., *Selected papers of Takeyuki Hida*, World Sci. Pub. Co., 2001.
2. S. Albeverio, A. Hahn, A. Sengupta, "Chern-Simons theory, Hida distributions, and state models", *Infinite Dim. Analysis, Quantum Probab. and Related Topics* **6**, 65–81 (2003).
3. S. Albeverio, A. Sengupta, "The Chern-Simon functional integral as an infinite distribution", *Nonlinear Analysis, Theory, Methods and Applications* **30**, 329–335 (1997).
4. A. Borodin, G. Olshanski, "Harmonic analysis on the infinite-dimensional unitary group and determinantal point processes", *Ann. Math.* **161**, 1329–1422 (2005).
5. I. M. Gel'fand, N. Ya. Vilenkin, *Generalized Functions*, vol. 4, *Applications of Harmonic Analysis*, Academic Analysis, Academic Press, 1964. Russian original, Moscow, 1961.
6. T. Hida, *Stationary Stochastic Processes*, Princeton Univ. Press, 1970.
7. T. Hida, *Brownian motion*, Springer-Verlag. 1980 (original Japanese edition, 1975).
8. T. Hida, H. Nomoto, "Gaussian measure on the projective limit of spheres", *Proc. Japan Acad.* **40**, 301–304 (1964).
9. T. Hida, Si Si, *Innovation Approach to Random Fields: An Application of White Noise Theory*, World Sci., 2004.
10. T. Hida, Si Si, *Lectures on White Noise Functionals*, World Sci. Pub. Co., 2006.
11. P. Lévy, *Processus stochastiques et mouvement brownien*, Gauthier-Villars, 1948. 2ème ed., 1965.
12. N. Obata, "Certain unitary representations of the symmetric group I", "II", *Nagoya Math. J.* **105**, 109–119 (1987); ibid. **106**, 143–162 (1987).
13. I. Ojima, "Macro-Micro duality in quantum physics", *Stochastic Analysis: Classical and Quantum. Perspectives of White Noise Theory*, ed. T. Hida, World Sci. Pub. Co., 2005, pp. 143–161.
14. S. Rohde, O. Schramm, "Basic properties of SLE", *Ann. Math.* **161**, 883–924 (2005).
15. Si Si, "Effective determination of Poisson noise", *Infinite Dim. Analysis and Quantum Probab. and Related Topics* **6**, 609–617 (2003).
16. Si Si, "Note on Poisson noise", *Quantum Information and Complexity*, eds. Hida, Saitô and Si Si, World Sci., 2004, pp. 411–425.
17. Si Si, A. Tsoi, Win Win Htay, "Invariance of Poisson noise", *Stochastic Analysis: Classical and Quantum*, ed. T.Hida, World Sci. Pub Co., 2005, pp. 199–210.

TAKEYUKI HIDA
Meijo University, Nagoya, Japan

SI SI
Aichi Prefectural University, Nagakute, Aichi-ken, Japan

ANALYSIS OF STABLE WHITE NOISE FUNCTIONALS

YUH-JIA LEE* (*National University of Kaohsiung*),
HSIN-HUNG SHIH† (*Kun Shan University, Tainan*)

The white noise analysis for Lévy processes developed in our previous paper [7] is extended to stable processes of order $\alpha \in (0,2)$. In the course of our investigation, the Segal–Bargmann transform for stable processes is introduced.

1. Introduction

Let $X = \{X(t) : t \in \mathbb{R}\}$ be a Lévy process, which is an additive process with stationary increments on a probability space $(\Omega, \mathcal{F}, P)$ with $X(0) = 0$ almost surely. By the well-known Lévy–Khintchine formula,

$$\mathbb{E}\big[e^{\mathrm{i}\, r(X(t)-X(s))}\big] = \exp[(t-s)\, f_X(r)] \tag{1}$$

for $r \in \mathbb{R}$ and $-\infty < s < t < +\infty$, where f_X is called the Lévy function of X which is uniquely expressed by

$$f_X(r) = \mathrm{i}\,\mu r - \frac{\sigma^2 r^2}{2} + \int_{|u|>0} \left(e^{\mathrm{i}\, ur} - 1 - \frac{\mathrm{i}\, ur}{1+u^2}\right) \frac{1+u^2}{u^2}\, d\beta(u), \quad r \in \mathbb{R},$$

where μ is a real constant, and β is a positive finite measure on $\mathbb{R}$ with $\sigma^2 = \beta(\{0\})$. Conversely, for a given triple (μ, σ, β), where $\mu, \sigma \in \mathbb{R}$ and β is a positive finite measure on $\mathbb{R}$, there exists a Lévy process X such that the equality (1) holds. X is called a Lévy process with the generating triple (μ, σ, β) and the measure $d\beta_0(u) = (1+u^2)/u^2\, d\beta(u)$ on $\mathbb{R} \setminus \{0\}$ is called the Lévy measure of X.

In our previous paper [7], under the assumption that β has finite absolute second moments, we have shown that the Lévy white noise measure Λ exists and defined as a Borel measure on the space $\mathcal{S}'$ of tempered distributions. Then the analysis of Lévy white noise functional were studied based on the following representation of a Lévy process $X(t)$:

$$X(t\,;x) = \begin{cases} \langle x,\, \mathbf{1}_{[0,\,t]}\rangle, & \text{if } t \geq 0, \\ -\langle x,\, \mathbf{1}_{[t,\,0]}\rangle, & \text{if } t < 0,\ x \in \mathcal{S}'. \end{cases} \tag{2}$$

*Research supported by the National Science Council of Taiwan
†Research supported by the National Science Council of Taiwan

Under the above assumption on the moments of β, Gaussian, Poisson, Gamma processes and Meixner are included in our investigation, but the stable processes with characteristic exponent $\alpha \in (0,2)$ are excluded since, in this case, the corresponding measure β has no second moments. Thus we wish to extend the theory developed in [7] to all α-stable processes. Let $0 < \alpha < 2$. A Lévy process $X = \{X(t) : t \in \mathbb{R}\}$ with $X(0) = 0$ a.s. is called an α-stable process on a probability space $(\Omega, \mathcal{F}, P)$ if its Lévy function has the following representation:

There are real constants μ, c_1, c_2 with $c_1, c_2 \geq 0$ and $c_1 + c_2 > 0$ such that for any $r \in \mathbb{R}$,

$$f_X(r) =$$
$$\mathrm{i}\,\mu\, r + \int\limits_{|u|>0} \left(e^{\mathrm{i}\,u r} - 1 - \frac{\mathrm{i}\,u\, r}{1+u^2}\right)\left(c_1\,\mathbf{1}_{(-\infty,0)}(u) + c_2\,\mathbf{1}_{(0,+\infty)}(u)\right)|u|^{-1-\alpha}\,du.$$

In other words, a stable process is a Lévy process with $\sigma = 0$ and β being given by

$$\beta_0(u) = \left(c_1\,\mathbf{1}_{(-\infty,0)}(u) + c_2\,\mathbf{1}_{(0,+\infty)}(u)\right)|u|^{-1-\alpha}. \tag{3}$$

Lemma 1.1 ([11, Theorem 14.15]). *Let $0 < \alpha < 2$ and X be an α-stable process on $(\Omega, \mathcal{F}, P)$. Then there are constants c, ρ with $c > 0$ and $\rho \in [-1, 1]$ so that*

$$f_X(r) = \mathrm{i}\,\mu\, r - c\,|r|^{\alpha}\left(1 + \mathrm{i}\,\rho\,\frac{r}{|r|}\,\omega(r,\alpha)\right), \qquad r \neq 0, \tag{4}$$

where $\omega(r,\alpha) = \tan\frac{\pi\alpha}{2}$ for $\alpha \neq 1$ and $\omega(r,\alpha) = \frac{2}{\pi}\ln|r|$ for $\alpha = 1$. Here, $\mu, c,$ and ρ are uniquely determined by $P \circ X(1)^{-1}$. Conversely, for any $c > 0$ and $\rho \in [-1,1]$ and $\mu \in \mathbb{R}$, there is an α-stable process X with f_X satisfying (4).

2. α-stable white noise functionals

By Lemma 1.1, there is a constant $\kappa > 0$ such that for each $r \in \mathbb{R}$,

$$|f_X(r)| \leq \begin{cases} \kappa\,(|r| + |r|^{\alpha}), & \text{if } \alpha \neq 1, \\ \kappa\,(|r| + |r\ln|r||), & \text{if } \alpha = 1. \end{cases} \tag{5}$$

Let Θ be the class given by $\Theta = L^1 \cap L^{\alpha}(\mathbb{R}, m)$ if $\alpha \neq 1$, and

$$\Theta = \left\{\phi \in L^1(\mathbb{R}, m);\ \int_{-\infty}^{+\infty} |\phi(t)\ln|\phi(t)||\,dt < +\infty\right\}$$

if $\alpha = 1$, where m is Lebesgue measure on $\mathbb{R}$. $L^1 \cap L^\alpha(\mathbb{R}, m)$, $0 < \alpha < 2$, is regarded as a metric space with the metric d_α defined by

$$d_\alpha(\phi, \psi) = \max\{|\phi - \psi|_{L^1(\mathbb{R},m)}, |\phi - \psi|^{\alpha\wedge 1}_{L^\alpha(\mathbb{R},m)}\}.$$

By (5) $f_X(\phi)$ is Lebesgue integrable for each $\phi \in \Theta$.

Lemma 2.1. *Let ϕ_n, $n \in \mathbb{N}$, and ϕ be elements in the class Θ which satisfy*

(*i*) *$\{\phi_n\}$ converges in measure to ϕ (with respect to m); and*

(*ii*) $\lim_{n\to\infty} \int_{-\infty}^{+\infty} (|\phi_n(t)| + |\phi_n(t)|^\alpha)\, dt = \int_{-\infty}^{+\infty} (|\phi(t)| + |\phi(t)|^\alpha)\, dt$ *for $\alpha \neq 1$;*

(*iii*) $\lim_{n\to\infty} \int_{-\infty}^{+\infty} |\phi_n(t)|(1 + |\ln|\phi_n(t)||)\, dt = \int_{-\infty}^{+\infty} |\phi(t)|(1 + |\ln|\phi(t)||)\, dt$ *for $\alpha = 1$.*

Then

$$\lim_{n\to\infty} \int_{-\infty}^{+\infty} f_X(\phi_n(t))\, dt = \int_{-\infty}^{+\infty} f_X(\phi(t))\, dt.$$

Proof. (Sketch)

— For any continuous function g on $\mathbb{R}$, it is not hard to prove that $\{g(\phi_n)\}$ converges in measure to $g(\phi)$.

— Applying the preceding result to $g(t) = |t|^\alpha$ for $0 < \alpha < 2$, $\alpha \neq 1$ and $g(t) = |t \ln|t||$ for $\alpha = 1$, the estimation (5) and the Lebesgue dominated convergence theorem, we obtain immediately

$$\int_{-\infty}^{+\infty} f_X(\phi_n(t))\, dt \to \int_{-\infty}^{+\infty} f_X(\phi(t))\, dt.$$

□

Let $(\cdot, \cdot)$ stand for the $\mathcal{S}'$-$\mathcal{S}$ pairing and observe that $\mathcal{S} \subset \Theta$. Consider the complex-valued functional $\mathcal{C}_X$ on $\mathcal{S}$ by

$$\mathcal{C}_X(\eta) = \exp\biggl(\int_{-\infty}^{+\infty} f_X(\eta(t))\, dt\biggr), \quad \eta \in \mathcal{S}. \tag{6}$$

Clearly $\mathcal{C}(0) = 1$ and $\mathcal{C}_X$ is a continuous positive-definite functional on the space $\mathcal{S}$ (by [1, p. 278]). It follows from Minlos's theorem, there exists a unique probability measure Λ on $(\mathcal{S}', \mathcal{B}(\mathcal{S}'))$, such that

$$\mathcal{C}_X(\eta) = \int_{\mathcal{S}'} e^{\mathrm{i}\,(x,\eta)}\, \Lambda(dx), \quad \eta \in \mathcal{S},$$

where $\mathcal{B}(\mathcal{S}')$ is the Borel field which is exactly the σ-field generated by all cylinder sets. Λ will be called the α-stable white noise measure.

Theorem 2.1. [9] *Let ϕ be an element in the class Θ. Then there is a sequence $\{\phi_n\}$ of elements of the space $\mathcal{S}$ so that (i) ϕ_n converges to ϕ in $L^1 \cap L^\alpha(\mathbb{R}, m)$, and (ii) $\{(\cdot, \phi_n)\}$ converges in probability to a random variable, denoted by $\langle \cdot, \phi \rangle$, with $\mathbb{E}[e^{\mathrm{i}\, r\, \langle \cdot, \phi\rangle}] = \exp\big(\int_{-\infty}^{+\infty} f_X(r\phi(t))\, dt\big)$, $r \in \mathbb{R}$.*

Remark 2.1.

1. In Theorem 2.1, ϕ_n's can be chosen so that $\phi_n \in \mathcal{K}$, the space of infinitely differentiable functions with compact supports.
2. It $\phi \in \mathcal{S}$, then $\langle \cdot, \phi \rangle = (\cdot, \phi)$ $[\Lambda]$-a.e..

Corollary 2.1. *For each $\phi \in \Theta$ with $\phi \neq 0$ $[m]$-a.e., the random variable $\langle \cdot, \phi \rangle$ has an α-stable distribution, where the corresponding Lévy function has the form as given in (4) associated with the triple (μ, c, ρ) of constants by*

$$\left(\mu \int_{-\infty}^{+\infty} \phi(t)\, dt,\ c \int_{-\infty}^{+\infty} |\phi(t)|^\alpha\, dt,\ \rho\, \frac{\int_{\{\phi>0\}} |\phi(t)|^\alpha\, dt - \int_{\{\phi<0\}} |\phi(t)|^\alpha\, dt}{\int_{-\infty}^{+\infty} |\phi(t)|^\alpha\, dt}\right)$$

or

$$\left(\mu \int_{-\infty}^{+\infty} \phi(t)\, dt - \frac{2c\rho}{\pi} \int_{-\infty}^{+\infty} \phi(t)\, \ln|\phi(t)|\, dt,\ c \int_{-\infty}^{+\infty} |\phi(t)|\, dt,\ \rho\, \frac{\int_{-\infty}^{+\infty} \phi(t)\, dt}{\int_{-\infty}^{+\infty} |\phi(t)|\, dt}\right),$$

according to $\alpha \neq 1$ or $\alpha = 1$.

Proof. Applying (4) to $\int_{-\infty}^{+\infty} f_X(r\phi(t))\, dt$, $r \in \mathbb{R}$, the proof is straightforward. □

Note that, for any bounded interval A, the indicator function $\mathbf{1}_A$ of A belongs to the class Θ. By Corollary 2.1, the α-stable process X on $(\mathcal{S}', \mathcal{B}(\mathcal{S}'), \Lambda)$ can be represented by:

$$X(t\,;\, x) = \begin{cases} \langle x,\, \mathbf{1}_{[0,\, t]} \rangle, & \text{if } t \geq 0, \\ -\langle x,\, \mathbf{1}_{[t,\, 0]} \rangle, & \text{if } t < 0,\ x \in \mathcal{S}'. \end{cases}$$

Proposition 2.1. *Let $\{a_n\}, \{b_n\}$ be two sequences of real numbers with $a_n \searrow -\infty$ and $b_n \nearrow +\infty$. Then for each $\phi \in \mathcal{S}$,*

$$\int_{a_n^+}^{b_n} \phi(t)\, dX(t; x) \longrightarrow (x, \phi)$$

for $[\Lambda]$-almost all $x \in \mathcal{S}'$ and the above integral is understood to be a Riemann–Stieltjes integral.

It follows from Proposition 2.1 that we may formally interpret the integral $\int_{-\infty}^{+\infty} x(t)\,\phi(t)\,dt$ by

$$\int_{-\infty}^{+\infty} x(t)\,\phi(t)\,dt = (x,\phi) = \int_{-\infty}^{+\infty} \phi(t)\,dX(t)$$
$$= \int_{-\infty}^{+\infty} \dot{X}(t;x)\,\phi(t)\,dt, \quad x \in \mathcal{S}'.$$

From the above interpretation we regard elements of $\mathcal{S}'$ as the sample path so that $\dot{X}(t;x) = x(t)$. Members of $L^2(\mathcal{S}',\Lambda)$ are referred as square-integrable α-stable white noise functionals.

Remark 2.2. If $\phi \in \Theta \setminus \mathcal{S}$, we define $\int_{-\infty}^{+\infty} \phi(t)\,dX(t)$ as the limit of convergence in probability of $\int_{-\infty}^{+\infty} \phi_n(t)\,dX(t)$, $n \in \mathbb{N}$, where $\{\phi_n\} \subset \mathcal{K}$ is a sequence satisfying the conditions (i) and (ii) stated in Theorem 2.1.

3. Chaos decomposition of α-stable white noise functionals

Let $\mathcal{B}_b(\mathbb{R}^2_*)$ be the class of all bounded Borel subsets E of $\mathbb{R}^2_* = \mathbb{R}^2 \setminus \{(t,0) : t \in \mathbb{R}\}$, away from the t-axis.

For $E \in \mathcal{B}_b(\mathbb{R}^2_*)$, let $N(E;\cdot)$ be a random variable on $(\mathcal{S}',\mathcal{B}(\mathcal{S}'))$ defined by

$$N(E;x) = \big|\{(t,u) \in E : X(t;x) - X(t-;x) = u\}\big|, \quad x \in \mathcal{S}'.$$

Then $N(E;x)$, $E \in \mathcal{B}_b(\mathbb{R}^2_*)$ and $x \in \mathcal{S}'$, is a Poisson random measure with the intensity measure ν, where

$$d\nu(t,u) = \big(c_1\,\mathbf{1}_{(-\infty,0)}(u) + c_2\,\mathbf{1}_{(0,+\infty)}(u)\big)|u|^{-1-\alpha}\,du\,dt.$$

Let

$$M(E) = \int_{\mathbb{R}^2_*} u\,\mathbf{1}_E(t,u)\,dN_0(t,u), \quad E \in \mathcal{B}(\mathbb{R}^2_*),$$

where $dN_0(t,u) = dN(t,u) - d\nu(t,u)$. Then $M(E)$, $E \in \mathcal{B}(\mathbb{R}^2_*)$, is an independent random measure with zero mean and

$$\mathbb{E}[M(E)\,M(F)] = \lambda(E \cap F)$$

for any $E, F \in \mathcal{B}(\mathbb{R}^2_*)$, where $d\lambda(t,u) = u^2\,d\nu(t,u)$. Let $I_n(g_n)$, $g_n \in \widehat{L^2_c}((\mathbb{R}^2_*)^n,\lambda^{\otimes n})$, be the nth order multiple integral of the kernel function g_n with respect to M, where $\widehat{L^2_c}((\mathbb{R}^2_*)^n,\lambda^{\otimes n})$ denotes the space of symmetric complex-valued L^2-functions on $(\mathbb{R}^2_*)^n$ with respect to $\lambda^{\otimes n}$. Then we have the isometry

$$\|I_n(g_n)\|^2_{L^2(\mathcal{S}',\Lambda)} = n! \int_{\mathbb{R}^2_*} \cdots \int_{\mathbb{R}^2_*} |g_n(s_1, \ldots, s_n)|^2 \, d\lambda(s_1) \cdots d\lambda(s_n).$$

The following chaos decomposition theorem is in fact a reformulation of [3, Theorem 2], due to K. Itô, on the probability space $(\mathcal{S}', \mathcal{B}(\mathcal{S}'), \Lambda)$ for stable processes.

Theorem 3.1. *For any $\varphi \in L^2(\mathcal{S}', \Lambda)$, there exists a sequence of kernel functions $\phi_n \in \widehat{L^2_c}((\mathbb{R}^2_*)^n, \lambda^{\otimes n})$, $n \in \mathbb{N} \cup \{0\}$, such that φ can be expressed uniquely as an orthogonal direct sum*

$$\sum_{n=0}^{\infty} I_n(\phi_n)$$

of $I_n(\phi_n)$, $n = 1, 2, \ldots$. In notation, we write $\varphi \sim (\phi_n)$.

The well-known Lévy–Itô decomposition theorem for Lévy processes (see [11, Theorem 19.2]) asserts that there is a set $\Delta \in \mathcal{B}(\mathcal{S}')$ with $\Lambda(\Delta) = 1$ such that for any $x \in \Delta$ and $b > a$,

$$X(b; x) - X(a; x) = \mu(b-a) + \lim_{n\to\infty} \int_a^b \int_{\frac{1}{n}<|u|<n} \left(u \, dN(t, u; x) - \frac{u}{1+u^2} \, d\nu(t, u) \right),$$

where the limit is uniformly convergent in a, b on any bounded interval.

In the case of stable processes, we have

$$X(b) - X(a) = \lim_{n\to\infty} \left\{ I_1(\mathbf{1}_{(a,b]\times\{\frac{1}{n}<|u|<n\}}) + \left(\mu + (c_2 - c_1) \int_{\frac{1}{n}}^{n} \frac{u^{2-\alpha}}{1+u^2} \, du \right)(b-a) \right\}. \quad (7)$$

Denote by

$$\tau_n = \mu + (c_2 - c_1) \int_{\frac{1}{n}}^{n} \frac{u^{2-\alpha}}{1+u^2} \, du, \quad n \geq 2, \quad \text{and } \tau = \lim_{n\to\infty} \tau_n. \quad (8)$$

We note that τ exists if and only if $1 < \alpha < 2$. Unlike the cases in [7], $X(b) - X(a) - \tau(b-a)$ is not a member of $L^2(\mathcal{S}', \Lambda)$ even though τ exists. It will be defined as a generalized stable white noise functional (see Section 7).

4. The Segal–Bargmann transform of square-integrable α-stable white noise functionals

For an arbitrarily given $\varphi \in L^2(\mathcal{S}', \Lambda)$, the Segal–Bargmann (or the S-) transform $S\varphi$ of φ is a complex-valued functional on $L^2_c(\mathbb{R}^2_*, \lambda)$ defined by

$$S\varphi(g) = \int_{\mathcal{S}'} \varphi(x)\, \mathcal{E}_X(g)(x)\, \Lambda(dx), \quad g \in L^2_c(\mathbb{R}^2_*, \lambda),$$

where $\mathcal{E}_X(g) = \sum_{n=0}^{\infty} \frac{1}{n!} I_n(g^{\otimes n})$, called the coherent state functionals associated with g. If $\varphi \sim (\phi_n)$, then

$$S\varphi(g) = \sum_{n=0}^{\infty} \int_{\mathbb{R}^2_*} \cdots \int_{\mathbb{R}^2_*} \phi_n(s_1, \dots, s_n)\, g(s_1) \cdots g(s_n)\, d\lambda(s_1) \cdots d\lambda(s_n). \quad (9)$$

In [6], we derived a closed form of the Segal–Bargmann transform $S\varphi(g)$ for $\varphi \in L^2(\mathcal{S}', \Lambda)$, and g belonging to some dense subset of $L^2_c(\mathbb{R}^2_*, \lambda)$. There measure β in [6] was assumed to satisfy the moment condition. If one carefully goes through the proof one would found that the moment condition is superfluous except some minor change. The major difference is that some regularity properties, such as the analyticity of characteristic functionals and square-integrability of cylinder functions are no longer hold. For the sake of clarity we sketch the proof by showing the key steps as follows.

Let $\mathcal{G}$ be the class of all bounded $\mathcal{B}(\mathbb{R}^2_*)$-measurable complex-valued functions which are supported on a compact set of $\mathbb{R}^2_*$. Then, for any $g \in \mathcal{G}$, we have

$$\begin{aligned}\int_{\mathcal{S}'} e^{I_1(g)(x)}\, \Lambda(dx) &= \exp\Big(-\int_{\mathbb{R}^2_*} g^*(t,u)\, d\nu(t,u)\Big) \\ &\quad \times \int_{\mathcal{S}'} \exp\Big(\int_{\mathbb{R}^2_*} g^*(t,u)\, dN(t,u;x)\Big)\, \Lambda(dx) \\ &= \exp\Big(\int_{\mathbb{R}^2_*} \big(e^{g^*(t,u)} - 1 - g^*(t,u)\big)\, d\nu(t,u)\Big), \end{aligned} \quad (10)$$

where $g^*(t,u) = u\, g(t,u)$ for $(t,u) \in \mathbb{R}^2_*$.

For any complex-valued $\mathcal{B}(\mathbb{R}^2_*)$-measurable function g, let

$$\Phi_g = \frac{e^{g^*(t,u)} - 1}{u}, \quad (t,u) \in \mathbb{R}^2_*.$$

Then, by making use of the integral formula (10), we have

Theorem 4.1. [10] *Let $g \in \mathcal{G}$ and $g_n \in \widehat{L^2_c}((\mathbb{R}^2_*)^n, \lambda^{\otimes n})$. Then*

$$\int_{\mathcal{S}'} I_n(g_n)(x)\, e^{I_1(g)(x)}\, \Lambda(dx) = \mathbb{E}\big[e^{I_1(g)}\big] \int_{\mathbb{R}^2_*} \cdots \int_{\mathbb{R}^2_*} g_n(t_1, u_1; \ldots; t_n, u_n) \times \prod_{j=1}^{n} \Phi_g(t_j, u_j)\, d\lambda(t_1, u_1) \cdots d\lambda(t_n, u_n).$$

Corollary 4.1. [10] *Let $g_n \in \widehat{L^2_c}((\mathbb{R}^2_*)^n, \lambda^{\otimes n})$. Then, for any $\phi \in \Theta$,*

$$\int_{\mathcal{S}'} I_n(g_n)(x)\, e^{\mathrm{i}\,\langle x, \phi\rangle}\, \Lambda(dx) = \mathbb{E}[e^{\mathrm{i}\,\langle \cdot, \phi\rangle}] \int_{\mathbb{R}^2_*} \cdots \int_{\mathbb{R}^2_*} g_n(t_1, u_1; \ldots; t_n, u_n) \times \prod_{j=1}^{n} \Phi_{\mathrm{i}\,\phi\otimes \mathbf{1}_{\mathbb{R}_*}}(t_j, u_j)\, d\lambda(t_1, u_1) \cdots d\lambda(t_n, u_n).$$

Corollary 4.2. [10] *The class of all functions $\Phi_{\mathrm{i}\,\eta\otimes\mathbf{1}_{\mathbb{R}_*}}$, $\eta \in \mathcal{S}$, is a total subset of $L^2_c(\mathbb{R}^2_*, \lambda)$.*

Combining Corollary 4.1 with (9) we immediately have

Proposition 4.1. *For any $\varphi \in L^2(\mathcal{S}', \Lambda)$ and $\phi \in \Theta$,*

$$\int_{\mathcal{S}'} \varphi(x)\, e^{\mathrm{i}\,\langle x,\, \phi\rangle}\, \Lambda(dx) = \mathbb{E}[e^{\mathrm{i}\,\langle\cdot, \phi\rangle}] \cdot S\varphi(\Phi_{\mathrm{i}\,\phi\otimes\mathbf{1}_{\mathbb{R}_*}}).$$

Let Ψ be the closed subspace $\{g \in L^2_c(\mathbb{R}^2_*, \lambda); g^* \in L^1_c(\mathbb{R}^2_*, \nu)\}$ of $L^2_c(\mathbb{R}^2_*, \lambda)$. Let $g \in \Psi$ be fixed. By [6, Proposition 2.1] there is a set $\Delta_g \in \mathcal{B}(\mathcal{S}')$ with $\Lambda(\Delta_g) = 1$ such that

$$\sum_{t\in\mathbb{R}} |g^*(t, j_X(t; x))| < +\infty \quad \text{for any } x \in \Delta_g,$$

where $j_X(t;x) = X(t;x) - X(t-;x)$, $t \in \mathbb{R}$. It implies that $1 + g^*(t, j_X(t;x)) \neq 0$ except only for finitely many $t \in \mathbb{R}$ and the infinite product $\prod_{t\in\mathbb{R}}(1 + g^*(t, j_X(t;x))$ is absolutely convergent. Define a functional $\Upsilon_X(g)$ associated with g by

$$\Upsilon_X(g)(x) = \begin{cases} \exp\big(-\int_{\mathbb{R}^2_*} g^*(t,u)\, d\nu(t,u)\big) \prod_{t\in\mathbb{R}} (1 + g^*(t, j_X(t;x))), & x \in \Delta_g, \\ 0, & \text{otherwise.} \end{cases}$$

Then $\Upsilon_X(g)$, $g \in \Psi$, is an entire function belonging to $L^2(\mathcal{S}', \Lambda)$. Moreover, for any $x \in \mathcal{S}'$,

$$|\Upsilon_X(g)(x)| \leq \exp\left(\int_{\mathbb{R}^2_*} |g^*(t,u)|\, d\nu(t,u) + \int_{\mathbb{R}^2_*} \ln(1 + |g^*(t,u)|)\, dN(t,u;x)\right)$$

and

$$\|\Upsilon_X(g)\|_{L^2(\mathcal{S}',\Lambda)} \leq \exp\left(2\int_{\mathbb{R}^2_*} |g^*(t,u)|\, d\nu(t,u) + \frac{1}{2}\int_{\mathbb{R}^2_*} |g(t,u)|^2\, d\lambda(t,u)\right),$$
$$g \in \Psi.$$

Theorem 4.2. *For any* $\varphi \in L^2(\mathcal{S}', \Lambda)$ *and* $g \in \Psi$,

$$S\varphi(g) = \int_{\mathcal{S}'} \varphi(x)\, \Upsilon_X(g)(x)\, \Lambda(dx).$$

5. Test and generalized functionals

From now on, the parameter c_1, c_2 in (3) will be always assumed to be $c_1, c_2 > 0$.

A Gel'fand triple on $L^2(\mathbb{R}^2_*, \lambda)$

Let $A = -d^2/dt^2 + (1+t^2)$ be a densely defined self-adjoint operator on $L^2(\mathbb{R}, dt)$ and $\{h_n;\, n \in \mathbb{N}_0\}$ be eigenfunctions of A with corresponding eigenvalues $2n+2$, $n \in \mathbb{N}_0 (\equiv \mathbb{N} \cup \{0\})$, where h_n's are Hermite functions on $\mathbb{R}$. Then $\{h_n;\, n \in \mathbb{N}_0\}$ forms a complete orthonormal basis (CONS, in abbreviation) of $L^2(\mathbb{R}, m)$. For any $p \in \mathbb{R}$ and $\eta \in L^2(\mathbb{R}, m)$, define $|\eta|_p := |A^p \eta|_{L^2(\mathbb{R},m)}$ and let $\mathcal{S}_p$ be the completion of the class $\{\eta \in L^2(\mathbb{R}, m);\, |\eta|_p < +\infty\}$ with respect to $|\cdot|_p$-norm. Then $\mathcal{S}_p$ is a real separable Hilbert space and we have the continuous inclusions:

$$\mathcal{S} = \varprojlim_{p>0} \mathcal{S}_p \subset \mathcal{S}_p \subset \mathcal{S}_q \subset L^2(\mathbb{R}, m) \subset \mathcal{S}_{-q} \subset \mathcal{S}_{-p} \subset \mathcal{S}' = \varinjlim_{p>0} \mathcal{S}_{-p},$$
$$p > q \geq 0,$$

and

$$\mathcal{S} \subset L^2(\mathbb{R}, m) \subset \mathcal{S}'$$

forms a Gel'fand triple.

Next, consider the real Hilbert space $L^2(\mathbb{R}_*, \gamma)$, where $d\gamma(u) = w(u)\, du$ with

$$w(u) = (c_1\, \mathbf{1}_{(-\infty,0)}(u) + c_2\, \mathbf{1}_{(0,+\infty)}(u))|u|^{1-\alpha}, \quad u \in \mathbb{R}_*.$$

Let $\zeta_n(u) = h_n(u)/\sqrt{w(u)}$, $u \in \mathbb{R}_*$, $n \in \mathbb{N}_0$. Then $\{\zeta_n;\, n \in \mathbb{N}_0\}$ is a CONS of $L^2(\mathbb{R}_*, \gamma)$. Define a linear operator A_α densely defined on $L^2(\mathbb{R}_*, \gamma)$ by $A_\alpha \zeta_n = (2n+2)\zeta_n$ for $n \in \mathbb{N}_0$ and for each $p \geq 0$, let $\mathcal{E}_p$ be the set of all $\phi \in L^2(\mathbb{R}_*, \gamma)$ with $|A_\alpha^p \zeta|_{L^2(\mathbb{R}_*,\gamma)} < +\infty$, which is a real separable Hilbert space with the inner product $\langle \cdot, \cdot \rangle_{p,\alpha}$ given by

$$\langle \phi,\, \psi \rangle_{p,\alpha} = \int_{\mathbb{R}_*} A_\alpha^p\, \phi(u)\, A_\alpha^p\, \psi(u)\, d\gamma(u), \quad \phi, \psi \in \mathcal{E}_p.$$

We use the notation $|\cdot|_{p,\alpha}$ to denote the induced norm by $\langle \cdot, \cdot \rangle_{p,\alpha}$.

Denote by $\mathcal{E}_{-p}$ the dual of $\mathcal{E}_p$. Then $\mathcal{E}_{-p}$ is isometrically isomorphic to the completion of $L^2(\mathbb{R}_*, \gamma)$ with the inner product $\langle \cdot, \cdot \rangle_{-p,\alpha}$ and $|\cdot|_{-p,\alpha}$-norm by

$$\langle \phi,\, \psi \rangle_{-p,\alpha} = \int_{\mathbb{R}_*} A_\alpha^{-p}\, \phi(u)\, A_\alpha^{-p}\, \psi(u)\, d\gamma(u), \quad \phi, \psi \in L^2(\mathbb{R}_*, \gamma).$$

Let $\mathcal{E} = \varprojlim_{p>0} \mathcal{E}_p$. Then $\mathcal{E}$ is a nuclear space with the dual $\mathcal{E}' = \varinjlim_{p>0} \mathcal{E}_{-p}$ and

$$\mathcal{E} \subset L^2(\mathbb{R}_*, \gamma) \subset \mathcal{E}'$$

also forms a Gel'fand triple. There is a connection between $\mathcal{E}_p$ and $\mathcal{S}_p$, $p \geq 0$, as follows.

Proposition 5.1. *Let $\phi \in L^2(\mathbb{R}_*, \gamma)$. Then for each $p \geq 0$, $\phi \in \mathcal{E}_p$ if and only if there is an element η_ϕ of $\mathcal{S}_p$ such that $\sqrt{w} \cdot \phi = \eta_\phi$, $[m]$-a.e. on $\mathbb{R}_*$. Moreover, for a sequence $\{\phi_n\}$ in $\mathcal{E}_p$, $\phi_n \to \phi$ in $\mathcal{E}_p$ if and only if $\eta_{\phi_n} \to \eta_\phi$ in $\mathcal{S}_p$.*

Corollary 5.1. *Let $f \in \mathcal{E}_{-p}$, $p > 0$. Then there is an element η_f of $\mathcal{S}_{-p}$ such that for every $\phi \in \mathcal{E}$,*

$$(f,\, \phi)_{\mathcal{E}',\mathcal{E}} = (\eta_f,\, \eta_\phi).$$

Moreover, if f is a regular distribution defined by a $\mathcal{B}(\mathbb{R}_*)$-measurable function, still denoted by f, that is, $(f, \phi)_{\mathcal{E}',\mathcal{E}} = \int_{\mathbb{R}_*} f(u)\, \phi(u)\, d\gamma(u)$, then $\eta_f = \sqrt{w} \cdot f$, $[m]$-a.e. on $\mathbb{R}_*$.

Remark 5.1. A crucial difference between the above argument and [7] is that in [7] we can apply the Gram–Schmidt orthogonalization method to $\{1, u, u^2, \ldots\}$ to obtain a CONS of $L^2(\mathbb{R}_*, \gamma)$ as in [7]. However in the case of a stable process, none of any polynomial functions $p(u)$, $u \in \mathbb{R}_*$, are square-integrable with respect to γ without the moment condition so that we can not apply the Gram–Schmidt orthogonalization method anymore.

Example 5.1.

1. **All polynomial functions $p(u)$, $u \in \mathbf{R}_*$, are regular elements in $\mathcal{E}'$.**
   In fact, let $\{\phi_n\} \subset \mathcal{E}$ such that $\phi_n \to \phi$ in $\mathcal{E}$. Then, by Proposition 5.1, $\eta_{\phi_n} \to \eta_\phi$ in $\mathcal{S}$ and

$$\begin{aligned}
&|\sqrt{w(u)}\,\eta_{\phi_n}(u)\,p(u)| \\
&\quad = (\sqrt{c_1}\,\mathbf{1}_{(-\infty,0)}(u) + \sqrt{c_2}\,\mathbf{1}_{(0,+\infty)}(u))|\eta_{\phi_n}(u)\,p(u)|\,|u|^{\frac{1}{2}-\frac{\alpha}{2}} \\
&\quad \leq \begin{cases} (\sqrt{c_1}+\sqrt{c_2})\,|\eta_{\phi_n}(u)\,p(u)|\,|u|^{\frac{1}{2}-\frac{\alpha}{2}}, & \text{if } 0<|u|<1, \\ (\sqrt{c_1}+\sqrt{c_2})\,|\,u^2\,\eta_{\phi_n}(u)\,p(u)|\,|u|^{-\frac{3}{2}-\frac{\alpha}{2}}, & \text{if } |u|\geq 1, \end{cases} \\
&\quad \leq \begin{cases} c\cdot |u|^{\frac{1}{2}-\frac{\alpha}{2}}, & \text{if } 0<|u|<1, \\ c\cdot |u|^{-\frac{3}{2}-\frac{\alpha}{2}}, & \text{if } |u|\geq 1, \end{cases}
\end{aligned}$$

   where $c > 0$ is a constant not depending on n. By the dominated convergence theorem,

$$\int_{\mathbb{R}_*} \sqrt{w(u)}\,\eta_{\phi_n}(u)\,p(u)\,du \to \int_{\mathbb{R}_*} \sqrt{w(u)}\,\eta_\phi(u)\,p(u)\,du$$

   and hence $p(u) \in \mathcal{E}'$.
2. **$\delta_u \in \mathcal{E}'$ for any $u \in \mathbf{R}_*$.**

Next, for $p \in \mathbb{R}$, denote by $\mathcal{N}_p$ the Hilbert space tensor product $\mathcal{S}_p \otimes \mathcal{E}_p$ with $\|\cdot\|_p$-norm defined by $\|h_n \otimes \zeta_m\|_p = |h_n|_p\,|\zeta_m|_{p,\alpha}$, and let $\mathcal{N} = \mathcal{S} \otimes \mathcal{E}$ with the dual $\mathcal{N}' = \mathcal{S}' \otimes \mathcal{E}'$. Then $\mathcal{N} \subset L^2(\mathbb{R}^2_*, \lambda) \subset \mathcal{N}'$ forms a Gel'fand triple and we have a continuous chain $\mathcal{N} \subset \mathcal{N}_p \subset \mathcal{N}_q \subset L^2(\mathbb{R}^2_*, \lambda) \subset \mathcal{N}_{-q} \subset \mathcal{N}_{-p} \subset \mathcal{N}'$, $p > q \geq 0$. In the following, we denote

$$\|\cdot\|_p := |\cdot|_{\mathcal{N}_{p,c}^{\widehat{\otimes} n}},$$

where $\widehat{\otimes}$ denotes the symmetric tensor product. Also, we relabel the CONS $\{h_n \otimes \zeta_m;\ n, m \in \mathbb{N}_0\}$ by $\{f_0, f_1, \ldots\}$, $f_0 = h_0 \otimes \zeta_0$, and let $\lambda_j = \|(A \otimes A_\alpha) f_j\|_0$ for any $j \in \mathbb{N}_0$.

Applying Proposition 5.1 and Corollary 5.1 we have the following

Proposition 5.2.

(*i*) *$\mathcal{N}_c$ is the space consisting of all functions g on $\mathbb{R}^2_*$ of the form given by*

$$g(t,u) = \frac{T_g(t,u)}{\sqrt{w(u)}}, \quad (t,u) \in \mathbb{R}^2_*,$$

where $T_g \in \mathcal{S}_c \otimes \mathcal{S}_c$. *Moreover, for a sequence* $\{g_n\}$ *and* g *in* $\mathcal{N}_c$, $g_n \to g$ *in* $\mathcal{N}_c$ *if and only if* $T_{g_n} \to T_g$ *in* $\mathcal{S}_c \otimes \mathcal{S}_c$.

(*ii*) *Let* $F \in \mathcal{N}_c'$. *There is an element* T_F *in* $\mathcal{S}_c' \otimes \mathcal{S}_c'$ *such that for every* $g \in \mathcal{N}_c$,

$$(F,\, g)_{\mathcal{N}_c',\mathcal{N}_c} = (T_F,\, T_g)_{\mathcal{S}_c'^{\otimes 2},\mathcal{S}_c^{\otimes 2}}.$$

Moreover, if F *is regular, then*

$$T_F(t,u) = \sqrt{w(u)} \cdot F(t,u), \quad [m^{\otimes 2}]\text{-}a.e.\ (t,u) \in \mathbb{R}_*^2.$$

(*iii*) *Let* $\mathcal{D}$ *be the class consisting of all complex-valued functions* $g(t,u)$, $(t,u) \in \mathbb{R}_*^2$, *in* $\mathcal{S}_c \otimes \mathcal{S}_c$, *which have compact supports away from the* t*-axis. Then* $\mathcal{D}$ *is dense in* $\mathcal{N}_c$.

Remark 5.2.

1. For each $g \in \mathcal{N}_c$, the function $\int_{\mathbb{R}_*} T_g(t,u)\sqrt{w(u)}\,du$, $t \in \mathbb{R}$, is an element in $\mathcal{S}_c$.
2. Each element f in $\mathcal{S}_c'$ can be identified with an element in $\mathcal{N}_c'$ by regarding f as $f \otimes \mathbf{1}_{\mathbb{R}_*}$. More precisely, for each $g \in \mathcal{N}_c$,

$$(f,\, g)_{\mathcal{N}_c',\mathcal{N}_c} = (f \otimes \mathbf{1}_{\mathbb{R}_*},\, g)_{\mathcal{N}_c',\mathcal{N}_c} = \left(f,\, \int_{\mathbb{R}_*} T_g(\cdot,u)\sqrt{w(u)}\,du\right).$$

Construction of test and generalized α-stable white noise functionals

Spaces of test and generalized α-stable white noise functionals will be constructed by using the second quantization operator $\Gamma((A \otimes A_\alpha)^p)$ of $(A \otimes A_\alpha)^p$, $p \in \mathbb{R}$, and applying the same procedure as in white noise analysis (see [5]). For $p \in \mathbb{R}$ and $\varphi \in L^2(\mathcal{S}', \Lambda)$, define

$$\|\varphi\|_p^2 = \|\Gamma((A \otimes A_\alpha)^p)\,\varphi\|_{L^2(\mathcal{S}',\Lambda)}^2.$$

and let $\mathcal{L}_p$ be the completion of the class $\{\varphi \in L^2(\mathcal{S}', \Lambda);\ \|\varphi\|_p < +\infty\}$ with respect to $\|\cdot\|_p$-norm. Then $\mathcal{L}_p$, $p \in \mathbb{R}$, is a Hilbert space with the inner product $\langle\!\langle \cdot, \cdot \rangle\!\rangle_p$ induced by $\|\cdot\|_p$-norm and we naturally come to the following facts. For the details, we refer the reader to [7].

Fact 5.1.

1. For $q - p > \frac{1}{2}$, $\mathcal{L}_q \subset \mathcal{L}_p$ and the embedding $\mathcal{L}_q \hookrightarrow \mathcal{L}_p$ is of Hilbert–Schmidt type.

2. Let $\mathcal{L} = \varprojlim_{p>0} \mathcal{L}_p$. Then $\mathcal{L}$ is a nuclear space with its dual $\mathcal{L}'$ and we have a continuous inclusion: $\mathcal{L} \subset \mathcal{L}_p \subset \mathcal{L}_q \subset \mathcal{L}_0 = L^2(\mathcal{S}', \Lambda) \subset \mathcal{L}_{-q} \subset \mathcal{L}_{-p} \subset \mathcal{L}' = \varinjlim_{p>0} \mathcal{L}_{-p}$, $p > q > 0$.
3. For $\varphi \sim (\phi_n) \in \mathcal{L}_p$ with $p > 0$, $\phi_n \in \mathcal{N}_{p,c}^{\widehat{\otimes} n}$ for each $n \in \mathbb{N}_0$ and $\|\varphi\|_p^2 = \sum_{n=0}^{\infty} n! |\phi_n|_p^2$.
4. Let $F \in \mathcal{L}_{-p}$, $p > 0$. Then there is a sequence $\{F_n\}$ with $F_n \in \mathcal{N}_{-p,c}^{\widehat{\otimes} n}$, $n \in \mathbb{N}_0$, such that for every $\varphi \sim (\phi_n) \in \mathcal{L}_p$,

$$\langle\langle F, \varphi \rangle\rangle = \sum_{n=0}^{\infty} n! \, (F_n, \phi_n)_{\mathcal{N}_{-p,c}^{\widehat{\otimes} n}, \mathcal{N}_{p,c}^{\widehat{\otimes} n}} \text{ and } \|F\|_{-p}^2 = \sum_{n=0}^{\infty} n! |F_n|_{-p}^2,$$

where $\langle\langle \cdot, \cdot \rangle\rangle$ denotes the $\mathcal{L}'$-$\mathcal{L}$ pairing, denoted by $F = \sum_{n=0}^{\infty} I_n(F_n)$ or in short, $F \sim (F_n)$.

$\mathcal{L}$ will serve as the space of test functions and the dual space $\mathcal{L}'$ of $\mathcal{L}$ the space of generalized functions, and members of $\mathcal{L}'$ are called generalized α-stable white noise functionals.

The S-transform on $L^2(\mathcal{S}', \Lambda)$ can be extended to all generalized α-stable white noise functionals as in [7] by

$$SF(g) = \langle\langle F, \mathcal{E}_X(g) \rangle\rangle, \quad F \in \mathcal{L}', \; g \in \mathcal{N}_c. \tag{11}$$

The following properties of SF can be derived directly from (11). See also [7,8] for details.

Proposition 5.3. *Let $F \sim (F_n)$ be in $\mathcal{L}_{-p}$, $p \in \mathbb{R}$. Then*

(*i*) *SF is an entire function in $\mathcal{N}_{p,c}$ satisfying*

$$|SF(g)| \leq \|F\|_{-p} \, e^{\frac{1}{2}|g|_p^2} \quad \text{for each } g \in \mathcal{N}_{p,c}.$$

(*ii*) *$D^n SF(0)(h_1, \ldots, h_n) = \langle\langle F, I_n(h_1 \widehat{\otimes} \cdots \widehat{\otimes} h_n) \rangle\rangle$ for any $h_1, \ldots, h_n \in \mathcal{N}_{p,c}$.*

(*iii*) *The S-transform is an unitary operator from $\mathcal{L}_{-p}$ onto the Bargmann–Segal–Dwyer space $\mathcal{F}^1(\mathcal{N}_{p,c})$ over $\mathcal{N}_{p,c}$. In fact,*

$$\|F\|_{-p}^2 = \sum_{n=0}^{\infty} \frac{1}{n!} \| D^n SF(0) \|_{HS^{(n)}(\mathcal{N}_{p,c})}^2,$$

where D^n is the nth Fréchet derivative of SF and $\|\cdot\|_{HS^{(n)}(H)}$ is the Hilbert–Schmidt operator norm of a n-linear functional on a Hilbert space H.

(*iv*) [8] *(Characterization theorem) Suppose that a complex-valued function G defined on $\mathcal{N}_c$ is analytic and satisfies the following growth condition:*

$$\exists\, p \in \mathbb{R},\ \exists\, c > 0 \ s.t.\ |G(\xi)| \le c e^{\frac{1}{2}|\xi|_{-p}^2}, \quad \forall\, \xi \in \mathcal{N}_c.$$

Then there exists a unique $F \in \mathcal{L}_{p-\frac{3}{2}}$ such that $SF = G$ with $\|F\|_{p-\frac{3}{2}} \le K c$, where K is a constant independent of the choice of F.

As an application of Proposition 5.3(iii), we have

Corollary 5.2. *For any $p \in \mathbb{R}$, the class $\{\mathcal{E}_X(g);\ g \in \mathcal{N}_c\}$ is a total subset of $\mathcal{L}_{-p}$.*

Proposition 5.4. *Let $\mathcal{D}$ be the class consisting of all functions $g(t,u)$, $(t,u) \in \mathbb{R}_*^2$, in $\mathcal{S}_c \otimes \mathcal{S}_c$, which have compact supports away from the t-axis.*

(*i*) *For each $g \in \mathcal{D}$, $\Phi_g \in \mathcal{N}_c$ and $e^{I_1(g)} \in \mathcal{L}$. Moreover, for each $F \in \mathcal{L}'$,*

$$\langle\langle F,\, e^{I_1(g)} \rangle\rangle = \mathbb{E}[e^{I_1(g)}] \cdot SF(\Phi_g).$$

(*ii*) *The set $\{e^{I_1(g)};\ g \in \mathcal{D}\}$ is a total subset of $\mathcal{L}$.*

Proof. (i) For a fixed $g \in \mathcal{D}$, it is obvious by Proposition 5.2 that Φ_g lies in $\mathcal{N}_c$. Now, for any $F \in \mathcal{L}'$, we take a sequence $\{\varphi_n\} \subset L^2(\mathcal{S}', \Lambda)$ and $\varphi_n \to F$ in $\mathcal{L}'$. By Theorem 4.1,

$$\lim_{n\to\infty} \int_{\mathcal{S}'} \varphi_n(x)\, e^{I_1(g)(x)}\, \Lambda(dx)$$
$$= \mathbb{E}[e^{I_1(g)}] \cdot \lim_{n\to\infty} S(\varphi_n)(\Phi_g) = \mathbb{E}[e^{I_1(g)}] \cdot SF(\Phi_g).$$

This implies that $e^{I_1(g)} \in \mathcal{L}$ and $\langle\langle F,\, e^{I_1(g)} \rangle\rangle = \mathbb{E}[e^{I_1(g)}] \cdot SF(\Phi_g)$.

(ii) It suffices to show that if $F \in \mathcal{L}'$ such that $SF(\Phi_g) = 0$ for all $g \in \mathcal{D}$, $F \equiv 0$. Since the function $r_g(t,u) = \log(1 + g^*(t,u))/u$, $(t,u) \in \mathbb{R}_*^2$, is in $\mathcal{D}$. Then for $g \in \mathcal{D}$ with $\mathrm{Re}(1 + g^*) > 0$, $SF(g) = SF(\Phi_{r_g}) = 0$. By Proposition 5.3, $SF \equiv 0$ on $\mathcal{D}$ and then by Proposition 5.2(iii), $SF \equiv 0$ on $\mathcal{N}_c$. Thus $F \equiv 0$. □

6. Annihilation, creation, and conservation operators

Annihilation and creation operators

Let $F \in \mathcal{L}_p$ and $\xi \in \mathcal{N}_{-p,c}$, $p \in \mathbb{R}$. The Gâteaux derivative

$$(d/dz)|_{z=0}\, SF(\cdot + z\,\xi)$$

in the direction ξ is an analytic function on $\mathcal{N}_{-p,c}$. In fact, by using the Cauchy integral formula, one can show that

$$(d/dz)|_{z=0}\, SF(\cdot + z\,\xi\,) \in \mathcal{L}_{p-1}.$$

Define

$$\partial_\xi\, F = S^{-1}\big((d/dz)|_{z=0}\, SF(\cdot + z\,\xi\,)\big).$$

Then, by Proposition 5.3(iv), we have $\partial_\xi\, F$ in $\mathcal{L}_{p-\frac{5}{2}}$.

Since ∂_ξ is continuous from $\mathcal{L}$ into itself, its adjoint operator ∂_ξ^* is then defined by

$$\langle\langle \partial_\xi^*\, F,\, \varphi\rangle\rangle := \langle\langle F,\, \partial_\xi\, \varphi\rangle\rangle \quad \text{for } F \in \mathcal{L}' \text{ and } \varphi \in \mathcal{L}.$$

∂_ξ is called the annihilation operator and ∂_ξ^* is called the creation operator.

Conservation operator

For every $p \geq 0$, denote by $\mathcal{M}_p$ the class consisting of all functions h in $\mathcal{N}_c$ so that the associated multiplication operator M_{h^*}, which is defined by $M_{h^*}(g) = h^*\, g$ for $g \in \mathcal{N}_{p,c}$, acts continuously from $\mathcal{N}_{p,c}$ into $L_c^2(\mathbb{R}^2, \lambda)$, where $h^*(t,u) = u\, h(t,u)$, $(t,u) \in \mathbb{R}^2$. For $h \in \mathcal{M}_p$, let $d\Gamma_h$ be the differential second quantization of M_{h^*}, and let ∂_h° be the linear operator on the linear space $\mathcal{H}$ spanned by $I_n(g_1 \widehat{\otimes} \cdots \widehat{\otimes}\, g_n)$, $g_1, \ldots, g_n \in \mathcal{N}_{p,c}$ and $n \in \mathbb{N}$, defined by

$$\partial_h^\circ\, I_n(g_1 \widehat{\otimes} \cdots \widehat{\otimes}\, g_n) = I_n(d\Gamma_h(g_1 \widehat{\otimes} \cdots \widehat{\otimes}\, g_n)).$$

Then, for $\varphi \in \mathcal{H}$ with $\varphi \sim (\phi_n)$,

$$\partial_h^\circ\, \varphi = \sum_{\text{finite}} n \int_{\mathbb{R}^2} \cdots \int_{\mathbb{R}^2} \text{sym}\{\phi_n(s_1, \ldots, s_n)\, h^*(s_n)\}\, dM(s_1) \cdots dM(s_n),$$

where "sym" means "the symmetrization of". Moreover,

$$\|\partial_h^\circ\, \varphi\|_0 \leq \|M_{h^*}\|\, \|\varphi\|_q, \quad q - p \geq 1,$$

where $\|M_{h^*}\|$ is the operator norm of M_{h^*}.

According the above estimation, we can extend the domain of ∂_h° to all $\varphi \in \mathcal{L}_p$. ∂_h° is called the conservation operator indexed by h.

Let $h \in \mathcal{M}_p$ with $p \geq 0$ and $g \in \mathcal{N}_c$. The Lévy product formula (see [7, Theorem 4.1]) still holds for α-stable processes as follows:

$$\begin{aligned} & I_1(h)\, I_m(g^{\otimes m}) \\ & \quad = m\langle h, g\rangle I_{m-1}(g^{\otimes m-1}) + I_{m+1}(h \,\widehat{\otimes}\, g^{\otimes m}) + m I_m(M_{h^*}(g) \widehat{\otimes}\, g^{\otimes m-1}) \\ & \quad = \partial_h\, I_m(g^{\otimes m}) + \partial_h^*\, I_m(g^{\otimes m}) + \partial_h^\circ\, I_m(g^{\otimes m}). \end{aligned}$$

In fact, by the same argument as in [7], we have

Theorem 6.1. *Let $h \in \mathcal{M}_p$ with $p \geq 0$ and $\varphi \sim (\phi_n) \in \mathcal{L}_q$ with $q - p \geq 1$. Then $\partial_h \varphi$, $\partial_h^* \varphi$, and $\partial_h^\circ \varphi$ are in $L_c^2(\mathcal{S}', \Lambda)$. Moreover, for $[\Lambda]$-almost all $x \in \mathcal{K}'$,*

$$I_1(h)(x)\, \varphi(x) = \partial_h\, \varphi(x) + \partial_h^*\, \varphi(x) + \partial_h^\circ\, \varphi(x). \tag{12}$$

7. A quantum decomposition of stable processes

Regarding the Lévy process $X(t;\, x)$ as a multiplicative operator acting on test Lévy white noise functionals, then $X(t;\, x)$ has a quantum decomposition provided that $\tau = \mu + \int_{-\infty}^{\infty} u\, d\beta(u)$ exists. If τ does not exist, we have the quantum decomposition for the renormalized Lévy process $X(t) - \tau t$. The former includes the cases such as Gaussian processes, Poisson processes, Gamma processes and the processes in the Meixner class; while the latter includes stable processes as special cases. This establishes a connection between Lévy white noise analysis and quantum probability theory.

Now, let $X(t; x)$, $t \in \mathbb{R}$ and $x \in \mathcal{S}'$, be a fixed α-stable process with $0 < \alpha < 2$. We need the following lemma for the further discussion.

Lemma 7.1. *There is a fixed $p_0 \in \mathbb{N}$ such that for any $\eta \in \mathcal{K}$, the space of infinitely differentiable functions on $\mathbb{R}$ with compact supports, the mapping from $\mathcal{N}_{p_0,c}$ into $L_c^2(\mathbb{R}_*^2, \lambda)$ by*

$$g(t,u) \mapsto u\, \eta(t) g(t,u), \quad (t,u) \in \mathbb{R}_*^2,$$

is continuous.

Proof. For any $g \in \mathcal{N}_c$, let $T_g(t,u)$, $(t,u) \in \mathbb{R}_*^2$, be defined as in Proposition 5.2(i). Then $T_g \in \mathcal{S}_c \otimes \mathcal{S}_c$. Let $p > 0$ such that $1 \otimes u^2 \in \mathcal{S}_{-p} \otimes \mathcal{S}_{-p}$ and let $q > p$ so that $|hk|_p \leq \text{Const.}\, |h|_q\, |k|_q$ for any $h, k \in \mathcal{S}_p \otimes \mathcal{S}_p$. Then, for $\eta \in \mathcal{K}$,

$$\begin{aligned}
&\int_{\mathbb{R}_*^2} u^2\,\eta(t)^2\,|g(t,u)|^2\,d\lambda(t,u)\\
&\qquad = \int_{\mathbb{R}^2} u^2\,\eta(t)^2\,|T_g(t,u)|^2\,du\,dt\\
&\qquad \le |\eta|_\infty^2\,|1\otimes u^2|_{-p}\cdot|T_g^2|_p\\
&\qquad \le \text{Const.}\,|\eta|_\infty^2\,|1\otimes u^2|_{-p}\,|T_g|_q^2\\
&\qquad = \text{Const.}\,|\eta|_\infty^2\,|1\otimes u^2|_{-p}\,\|g\|_q^2.
\end{aligned}$$

Let $p_0 = q$ and then we complete the proof. □

By Lemma 7.1, the definition of the conservation operator ∂_h° can be extended to $h \in \mathcal{K}\otimes 1$. For convenience, we write ∂_η° as $\partial_{\eta\otimes\mathbf{1}_{\mathbb{R}_*}}^\circ$ for $\eta \in \mathcal{K}$. Moreover, as in [7, Theorem 6.3] the conservation operator can also be written as follows: for $\eta \in \mathcal{K}$ and $\varphi \in \mathcal{L}$,

$$\partial_\eta^\circ\,\varphi = \int_{\mathbb{R}_*^2} u\eta(t)\,\partial_{\delta_{(t,u)}}^*\,\partial_{\delta_{(t,u)}}\,\varphi\,w(u)\,du\,dt \quad \text{in } \mathcal{L}', \tag{13}$$

where $\delta_{(t,u)} = \delta_t \otimes \delta_u$ and the integral exists in the sense of Bochner.

Proposition 7.1. *For $\varphi \in \mathcal{L}$ and $\phi \in \Theta$, let $\{\phi_n\} \subset \mathcal{K}$ be a sequence converging to ϕ in $L^1(\mathbb{R}, m)$. Then for any increasing sequence $\{A_n\}$ of compact subsets of $\mathbb{R}_*$ with $\lim_{n\to\infty} A_n = \mathbb{R}_*$,*

$$\mathcal{L}' - \lim_{n\to\infty} I_1(\phi_n \otimes \mathbf{1}_{A_n})\,\varphi$$

exists. We denote such a limit by $I_1(\phi)\,\varphi$.

Proof. (Sketch)

- $\{\partial_{\phi_n\otimes\mathbf{1}_{A_n}}\,\varphi\}$ and $\{\partial_{\phi_n\otimes\mathbf{1}_{A_n}}^*\,\varphi\}$ are convergent to $\partial_\phi\,\varphi$ and $\partial_\phi^*\,\varphi$ in $\mathcal{L}'$ respectively.
- By Lemma 7.1, it is not hard to prove that for $q - p_0 \ge 1$ and $\eta \in \mathcal{K}$,

$$\sum_m (2m+2)^{-2q} \int_{\mathbb{R}_*} |u|\,|\zeta_m(u)|^2\,w(u)\,du < +\infty.$$

Then, by the preceding result, we obtain that

$$\begin{aligned}&\left\| \int_{\mathbb{R}_*^2} u\,(\phi_n(t)\,\mathbf{1}_{A_n}(u) - \phi(t))\,\partial^*_{\delta_{(t,u)}}\partial_{\delta_{(t,u)}}\,\varphi\, w(u)\,du\,dt \right\|_{-q} \\ &\leq \text{Const.}\left\{ \left(\int_{-\infty}^{+\infty} |\phi_n(t) - \phi(t)|\,|\delta_t|_{-q}^2\,dt \right) \right. \\ &\qquad \times \left(\sum_m (2m+2)^{-2q} \int_{\mathbb{R}_*} |u|\,|\zeta_m(u)|^2\,w(u)\,du \right) \\ &\qquad + \left(\int_{-\infty}^{+\infty} |\phi(t)|\,|\delta_t|_{-q}^2\,dt \right) \\ &\qquad \left. \times \left(\sum_m (2m+2)^{-2q} \int_{\mathbb{R}_*\setminus A_n} |u|\,|\zeta_m(u)|^2\,w(u)\,du \right) \right\} \|\varphi\|_q \\ &\to 0 \quad \text{as } n \to \infty.\end{aligned}$$

— Apply the above estimation. Then, by (13) and Theorem 6.1 for $I_1(\phi_n)$, the proposition follows immediately.

□

For any real-valued random variable Y on $(\mathcal{S}', \mathcal{B}(\mathcal{S}'))$, let M_Y denote the multiplication operator by Y, i.e., $M_Y\varphi = Y\,\varphi$, $\varphi \in L^2(\mathcal{S}', \Lambda)$.

Proposition 7.2. *Let $\phi \in \Theta$. Then there exist sequences $\{\phi_n\}$ and $\{A_n\}$ with properties as in Proposition 7.1 such that*

$$\exp\!\left(\mathrm{i}\,t\,\tau_{A_n} \int_{-\infty}^{+\infty} \phi_n(t)\,dt + \mathrm{i}\,t\,M_{I_1(\phi_n \otimes \mathbf{1}_{A_n})} \right) \to \exp\!\left(\mathrm{i}\,t\,M_{\langle \cdot, \phi \rangle}\right)$$

in the sense of strong convergence for any $t \in \mathbb{R}$, where $\tau_{A_n} = \mu + \int_{A_n} u\,d\beta(u)$.

Proof. Let $Y_{n,t} = I_1((\mathrm{sgn}(t)\,\mathbf{1}_{[0\wedge t,\,0\vee t]}) \otimes \mathbf{1}_{B_n}) + \tau_n\,t$, $n \in \mathbb{N}$, $t \in \mathbb{R}$, where $B_n = \{u;\ 1/n \leq |u| \leq n\}$. By the Lévy–Itô decomposition theorem, there is a set $\Delta \in \mathcal{B}(\mathcal{S}')$ with $\Lambda(\Delta) = 1$ such that for any $\eta \in \mathcal{K}$ and $x \in \Delta$,

$$(x,\,\eta) = \int_{-\infty}^{+\infty} \eta(t)\,dX(t;x) = \lim_{n\to\infty} \int_{-\infty}^{+\infty} \eta(t)\,dY_{n,t}(x).$$

For ϕ, we take a sequence $\{\phi_m\} \subset \mathcal{K}$ with the property stated in Theorem 2.1. Then $(\cdot,\,\phi_m) \to \langle \cdot,\,\phi \rangle$ in probability as $m \to \infty$. Since $\int_{-\infty}^{+\infty} \phi_m(t)\,dY_{n,t}$ $\to (\cdot, \phi_m)$ in probability as $n \to \infty$, we can choose a subsequence $\{k_m\}$ of $\{n\}$ such that $\int_{-\infty}^{+\infty} \phi_m(t)\,dY_{k_m,t} \to \langle \cdot, \phi \rangle$ in probability as $m \to \infty$, and thus there is a subsequence $\{\ell_n\}$ of $\{m\}$ such that $Z_n \equiv \int_{-\infty}^{+\infty} \phi_{\ell_n}(t)\, dY_{k_{\ell_n},t} \to \langle \cdot, \phi\rangle$ $[\Lambda]$-almost surely. Therefore,

$$\int_{\mathcal{S}'} \left| e^{\mathrm{i}\, t\, Z_n(x)} - e^{\mathrm{i}\, t\, \langle x, \phi\rangle} \right|^2 |\varphi(x)|^2\, \Lambda(dx) \to 0, \quad \forall\, \varphi \in L^2(\mathcal{S}', \Lambda).$$

It implies that $e^{\mathrm{i}\, t\, M_{Z_n}} \to e^{\mathrm{i}\, t\, M_{\langle \cdot, \phi\rangle}}$ strongly for any $t \in \mathbb{R}$. Since

$$Z_n = \tau_{k_{\ell_n}} \int_{-\infty}^{+\infty} \phi_{\ell_n}(t)\, dt + \mathrm{i}\, t\, M_{I_1(\phi_{\ell_n} \otimes \mathbf{1}_{B_{\ell_n}})},$$

the proof is completed. □

Apply the above propositions for $\phi = \mathbf{1}_{[0,t]}$, $t \geq 0$; or $\phi = \mathbf{1}_{[t,0]}$, $t < 0$. It is natural to regard "$X(t) - \tau\, t$" as a generalized stable white noise functional by

$$\langle\!\langle X(t) - \tau\, t,\, \varphi\rangle\!\rangle = \lim_{n\to\infty} \langle\!\langle I_1(\phi_n \otimes \mathbf{1}_{A_n}),\, \varphi\rangle\!\rangle, \quad \varphi \in \mathcal{L}.$$

By Theorem 6.1 and Proposition 7.1, we can obtain the quantum decomposition of stable processes as follows.

Theorem 7.1. *The "renormalized" Lévy process $X(t) - \tau\, t$ is a continuous operator from $\mathcal{L}$ into $\mathcal{L}'$ and we have*

$$\begin{aligned}(X(t) - \tau\, t)\, \varphi = {} & \partial_{\mathbf{1}_{[0,t]}\otimes 1}\, \varphi + \partial^*_{\mathbf{1}_{[0,t]}\otimes 1}\, \varphi \\ & + \int_0^t \Big\{ c_1 \int_{-\infty}^0 u\, |u|^{1-\alpha}\, \partial^*_{\delta_{(t,u)}}\, \partial_{\delta_{(t,u)}}\, \varphi\, du \\ & \quad + c_2 \int_0^{+\infty} u^{2-\alpha}\, \partial^*_{\delta_{(t,u)}}\, \partial_{\delta_{(t,u)}}\, \varphi\, du \Big\}\, dt. \end{aligned} \tag{14}$$

If τ is finite, we obtain the quantum decomposition for $X(t)$ from the identity (14).

If we formally take the derivative in both sides of the identity (14) with respect to t we obtain

$$\int_{-\infty}^{+\infty} (\dot{X}(t; x) - \tau)\, \varphi(x))\, \eta(t)\, dt = \int_{-\infty}^{+\infty} \eta(t)\, \big(\partial_t + \partial_t^* + \partial_t^\circ\big)\varphi(x)\, dt, \quad \eta \in \mathcal{K},$$

where $\partial_t = \partial_{\delta_t}$, $\partial_t^* = \partial^*_{\delta_t}$, and $\partial_t^\circ = \partial^\circ_{\delta_t}$. Symbolically we may write $:\!\dot{X}(t)\!: = X(t) - t$ which is called a renormalization of $X(t)$. Then the quantum decomposition of the generalized functional $:\!\dot{X}(t)\!:$ is given by

$$:\!\dot{X}(t)\!: = \partial_t + \partial_t^* + \partial_t^\circ.$$

Acknowledgement

The final version of the present paper was completed while the first author was visiting CCM (Centro de Ciências Matemáticas, Universidade da Madeira) in March, 2006. The first author would like to thank the Universidade da Madeira for financial support while visiting CCM.

References

1. I. M. Gel'fand, N. Y. Vilenkin, *Generalized Functions*, vol. IV, Academic Press, New York, 1964.
2. T. Hida, *Analysis of Brownian Functionals*, *Carleton Mathematical Lecture Notes* no. 13, 2nd ed., 1978.
3. K. Itô, Spectral type of shift transformations of differential process with stationary increments, *Trans. Amer. Math. Soc.* **81**, 253–263 (1956).
4. Y. Ito, Generalized Poisson functionals, *Probab. Theory Relat. Fields* **77**, 1–28 (1988).
5. H.-H. Kuo, *White Noise Distribution Theory*, CRC Press, 1996.
6. Y.-J. Lee, H.-H. Shih, "The Segal–Bargmann transform for Lévy functionals", *J. Funct. Anal.* **168**, 46–83 (1999).
7. Y.-J. Lee, H.-H. Shih, "Analysis of generalized Lévy white noise functionals", *J. Funct. Anal.* **211**, 1–70 (2004).
8. Y.-J. Lee, H.-H. Shih, "A characterization of generalized Lévy white noise functionals", *Quantum Information and Complexity*, World scientific, 2004, pp. 321–339.
9. Y.-J. Lee, H.-H. Shih, "Lévy white noise measure on infinite dimensional spaces: existence and characterization of measurable support", *J. Funct. Anal.* (2006), in press.
10. Y.-J. Lee, H.-H. Shih,
11. K.-I. Sato, *Lévy Processes and Infinitely Divisible Distributions*, *Cambridge Studies in Advanced Mathematics*, Cambridge University Press, Cambridge, 1999.

YUH-JIA LEE
Department of Applied Mathematics, National University of Kaohsiung,
Kaohsiung, TAIWAN 811

HSIN-HUNG SHIH
Department of Accounting Information, Kun Shan University,
Tainan, TAIWAN 710

UNITARIZING MEASURES FOR A REPRESENTATION OF THE VIRASORO ALGEBRA, ACCORDING TO KIRILLOV AND MALLIAVIN: STATE OF THE PROBLEM

PAUL LESCOT (*INSSET, Saint-Quentin*)

0. Introduction

In this paper, we review results due to Airault-Bogachev ([1]), Airault-Malliavin ([2]) and Kirillov ([7],[8],[9]). In the first paragraph, we define the Virasoro algebra $\mathcal{V}ir_{c,h}$ (depending upon two real parameters $c > 0$ and $h \geq 0$) as a natural central extension of the (complexified algebra of) the Lie algebra $\text{diff}(S^1)$ of $\mathcal{C}^\infty$-diffeomorphisms of the unit circle S^1, and introduce a complex structure on its subspace $\text{diff}_0(S^1)$. In the second paragraph we explain the identification of the (infinite-dimensional) homogeneous space $\text{Diff}(S^1)/S^1$ with a certain space $\mathcal{M}$ of univalent functions on the (open) unit disk; this is due to Kirillov ([7], [8], [9]), as well as the definition of a doubly infinite sequence $(L_k)_{k\in\mathbb{Z}}$ of differential operators on $\mathcal{M}$, and its link to the Neretin polynomials; our exposition of these matters in §2 and the first half of §3 follows [2] and [9]. In the second half of §3 the previous ingredients are combined in order to define a representation ρ of the Virasoro algebra; this is due to Airault-Malliavin ([2]), and, in a more general case, to Airault-Malliavin-Thalmaier ([3]). To these authors is also due the notion of a *unitarizing measure* on $\mathcal{M}$ for ρ, that we define in §4; we conclude by giving a proof of the result from [1] that such a measure, should it exist, would admit the above-mentioned Neretin polynomials as an orthogonal family. This entails as a consequence the non-existence of a unitarizing measure in the case $h = 0$, obtained in [1] in an almost purely algebraic way, and previously in [2] by a geometrical argument.

The content of this paper is based upon a talk given at Bielefeld University on July 27th, 2004; I have tried to uniformize notation, sign conventions, etc., and I have slightly amended the definition of ρ given in [1]. A preliminary version appeared as a preprint in the Bielefeld BIBOS series in 2004; I am indebted to Professor Michael Roeckner for the invitation to lecture in Bielefeld upon these topics, and to Professors Ana Bela Cruzeiro and Jean-Claude Zambrini, resp. Carlos Florentino, for invitations to give talks in Lisbon upon this material, in April, resp. June, 2005. To Professors Hélène Airault, Philippe Blanchard and Francesco Russo, as well as to the anonymous referee, I address my thanks for numerous remarks on the first version of the paper.

My recent joint work ([5]) with Professor M. Gordina is an attempt at a new approach to some of the questions described here. Of related interest is the paper [4].

1. Preliminaries

Let $\mathrm{Diff}(S^1)$ denote the group of $\mathcal{C}^\infty$, orientation-preserving diffeomorphisms of the unit circle S^1. Its Lie algebra $\mathrm{diff}(S^1)$ can be naturally identified with the set of $\mathcal{C}^\infty$ vector fields on S^1, i.e.:

$$\mathrm{diff}(S^1) = \left\{ \phi(\theta)\frac{d}{d\theta} \;\middle|\; \phi : \mathbb{R} \to \mathbb{R},\, \mathcal{C}^\infty, 2\pi\text{-periodic} \right\}.$$

We shall often identify, without further warning, the function ϕ and the vector field $\phi(\theta)\frac{d}{d\theta}$. A topological basis (for the obvious Fréchet space topology) of $\mathrm{diff}(S^1)$ is given by the $(f_k)_{k\geq 0}$ and the $(g_k)_{k\geq 1}$, where:

$$f_k \stackrel{\text{def}}{=} \cos(k\theta)\frac{d}{d\theta}$$

and

$$g_k \stackrel{\text{def}}{=} \sin(k\theta)\frac{d}{d\theta}\,.$$

Let $\mathrm{diff}_{\mathbb{C}}(S^1) \stackrel{\text{def}}{=} \mathrm{diff}(S^1) \otimes_{\mathbb{R}} \mathbb{C}$ denote the complexified Lie algebra of $\mathrm{diff}(S^1)$; it is now clear that a topological basis of $\mathrm{diff}_{\mathbb{C}}(S^1)$ is given by the $\{e_k\}_{k\in\mathbb{Z}}$, where

$$e_k \stackrel{\text{def}}{=} e^{ik\theta}\frac{d}{d\theta}\,.$$

One has the commutation relations

$$[e_k, e_{k'}] = i(k' - k)e_{k+k'}\,.$$

The Lie algebra $\mathrm{diff}_{\mathbb{C}}(S^1)$ contains

$$\mathcal{A} \stackrel{\text{def}}{=} \mathrm{Vect}_{\mathbb{C}}(e_k)_{k\in\mathbb{Z}}$$

as a Lie subalgebra, dense in the natural Fréchet space topology.

Setting $L_k = -ie_k$, one finds that:

$$[L_k, L_{k'}] = (k' - k)L_{k+k'},$$

whence

$$\mathcal{A} \simeq \mathrm{Der}_{\mathbb{C}}(\mathbb{C}[t, t^{-1}]).$$

Here L_k corresponds, through this isomorphism, to $t^{k+1}\frac{d}{dt}$, which is equivalent to setting $t = e^{i\theta}$.

The algebra $\mathcal{V}ir_{c,h}$ is defined by:

$$\mathcal{V}ir_{c,h} \stackrel{\text{def}}{=} \mathrm{diff}_{\mathbb{C}}(S^1) \oplus \mathbb{C}\kappa$$

as a vector space, with the following Lie bracket for any $(f, g) \in \mathrm{diff}_{\mathbb{C}}(S^1)^2$,

$$[\kappa, f] = 0,$$

and

$$[f, g]_{\mathcal{V}ir_{c,h}} = [f, g] + \omega_{c,h}(f, g)\kappa,$$

where

$$\omega_{c,h}(f, g) \stackrel{\text{def}}{=} \int_0^{2\pi} \left(\left(2h - \frac{c}{12} \right) f'(\theta) - \frac{c}{12} f'''(\theta) \right) g(\theta) \frac{d\theta}{2\pi}.$$

The so-called *Gelfand-Fuks cocycle* is $\omega_{0,1}$.

It turns out that $\mathcal{V}ir_{c,h}$ is the unique nontrivial central extension of $\mathrm{diff}_{\mathbb{C}}(S^1)$ (see [6]). An easy computation yields

Proposition 1.1. *For any* $(m, n) \in \mathbb{Z}^2$

$$\omega_{c,h}(e_m, e_n) = i \left[2hm + c\frac{m^3 - m}{12} \right] \delta_{m,-n}.$$

It is easy to deduce from [6, chapter 7, exercises 7.1 and 7.13], that the $\omega_{c,h}$ are exactly the continuous cocycles α on $\mathrm{diff}_{\mathbb{C}}(S^1)$ such that for any f in $\mathrm{diff}_{\mathbb{C}}(S^1)$

$$\alpha(e_0, f) = 0.$$

We shall denote by $\mathcal{V}ir^{\mathbb{R}}_{c,h}$ the natural "real" Lie subalgebra of $\mathcal{V}ir_{c,h}$, i.e.

$$\mathcal{V}ir^{\mathbb{R}}_{c,h} \stackrel{\text{def}}{=} \mathrm{diff}(S^1) \oplus \mathbb{R}\kappa.$$

From now on, we shall assume $c > 0$ and $h \geq 0$. Let

$$\mathrm{diff}_0(S^1) \stackrel{\text{def}}{=} \left\{ \phi \in \mathrm{diff}(S^1) \;\middle|\; \int_0^{2\pi} \phi(\theta) d\theta = 0 \right\}.$$

On $\mathrm{diff}_0(S^1)$, one defines a complex structure as follows: for

$$\phi(\theta) = \sum_{k=1}^{+\infty} (a_k \cos(k\theta) + b_k \sin(k\theta)),$$

the sequences $\{a_k\}_{k\geq 0}$ and $\{b_k\}_{k\geq 1}$ are rapidly decreasing, i.e., for any $m \geq 0$ one has

$$n^m(|a_n| + |b_n|) \underset{n\to+\infty}{\longrightarrow} 0\,.$$

We now set

$$J\phi(\theta) \stackrel{\text{def}}{=} \sum_{k=1}^{+\infty} (-a_k \sin(k\theta) + b_k \cos(k\theta))\,,$$

and we have $J\phi \in \mathcal{C}^\infty(S^1)$.

Lemma 1.1 ([2, p. 630]). *For any f in* $\mathrm{diff}(S^1)$

$$\omega_{c,h}(f, Jf) = \frac{1}{2} \sum_{k=1}^{+\infty} \Big[2hk + \frac{c}{12}(k^3 - k)\Big]\, (a_k^2 + b_k^2) \geq 0\,.$$

Proof. Let us set $c_k(f) = \int_0^{2\pi} e^{-ik\theta} f(\theta) \frac{d\theta}{2\pi}$, then $f(\theta) = \sum_{k\in\mathbb{Z}} c_k e^{ik\theta}$ and

$$Jf(\theta) = \sum_{k\geq 1} (ic_k e^{ik\theta} - ic_{-k} e^{-ik\theta})$$

whence (by Proposition 1.1):

$$\begin{aligned} \omega_{c,h}(f, Jf) &= \sum_{k\geq 1} c_k(-ic_{-k}) \Big(i\Big(2h - \frac{c}{12}\Big)k + \frac{ic}{12}k^3\Big) \\ &\quad + \sum_{k\geq 1} c_{-k}(-ic_k) \Big(i\Big(2h - \frac{c}{12}\Big)(-k) - \frac{ic}{12}k^3\Big) \\ &= 2\sum_{k\geq 1} (c_k c_{-k}) \Big(2hk + \frac{c}{12}(k^3 - k)\Big)\,. \end{aligned}$$

Taking into account the relations: for any $k \geq 1$, $a_k = c_k + c_{-k}$ and $b_k = i(c_k - c_{-k})$, the result follows. □

2. Kirillov's construction of an action of Diff(S^1) on a space of univalent functions

Let $D = \mathcal{D}(0,1)$ denote the open unit disk in $\mathbb{C}$, and let $\mathcal{M}$ denote the set of $\mathcal{C}^\infty$ functions $f : \bar{D} \to \mathbb{C}$, injective, holomorphic on D, with $f(0) = 0$, $f'(0) = 1$, and $f'(z) \neq 0$ for any $z \in \bar{D}$. Each $f \in \mathcal{M}$ can be written as

$$f(z) = z\left(1 + \sum_{n=1}^{+\infty} c_n z^n\right), \quad z \in D,$$

whence the imbedding

$$\mathcal{M} \hookrightarrow \mathbb{C}^{\mathbb{N}^*}$$
$$f \mapsto (c_1, c_2, \ldots)\,.$$

In fact, by De Branges's solution of Bieberbach's conjecture, one has $|c_n| \leq n+1$, thus one may identify $\mathcal{M}$ with an open subset of $\Pi_{n\geq 1}\mathcal{B}_{\mathbb{C}}(0, n+2)$; one therefore obtains a structure of (contractible) manifold on $\mathcal{M}$.

For $f \in \mathcal{M}$, $\Gamma = f(S^1) = f(\partial D)$ is a Jordan curve, therefore one has a decomposition into connected components:

$$(\mathbb{C} \cup \{\infty\}) \setminus \Gamma = \Gamma^+ \cup \Gamma^-$$

with $0 \in \Gamma^+$ and $\infty \in \Gamma^-$. By a combination of Riemann's Representation Theorem and Caratheodory's Theorem, there exists a holomorphic mapping

$$\phi_f : (\mathbb{C} \cup \{\infty\} \setminus D) \to \overline{\Gamma^-} = \Gamma^- \cup \Gamma$$

such that $\phi_f(\infty) = \infty$. Let us then define g_f by

$$g_f : S^1 \to S^1$$
$$e^{i\theta} \mapsto f^{-1}(\phi_f(e^{i\theta}))\,.$$

Then $g_f \in \text{Diff}(S^1)$, and g_f is well-defined up to multiplication on the right by an holomorphic automorphism of $\mathbb{C} \setminus \bar{D}$ stabilizing ∞, i.e., a rotation, whence a mapping

$$\mathcal{K} : \mathcal{M} \to \text{Diff}(S^1)/S^1\,.$$

Theorem 2.1 (Kirillov, [9, p. 736]). *$\mathcal{K}$ is a bijection.*

Therefore, by transport of structure, $\text{Diff}(S^1)/S^1$ acquires a structure of contractible complex manifold. Using J and $\omega_{c,h}$, this manifold can be equipped with a Kählerian structure(see [2], [7]).

Definition 2.1 (Kirillov action). For $v = \phi(\theta)\frac{d}{d\theta} \in \operatorname{diff}(S^1)$ and $f \in \mathcal{M}$ let us write $w(e^{i\theta}) = \phi(\theta)$, and define $K_v(f) : D \to \mathbb{C}$ by the following equality, for any $z \in D$ (we set $t = e^{i\theta}$)

$$K_v(f)(z) = \frac{f(z)^2}{2\pi}\int_0^{2\pi}\left(\frac{tf'(t)}{f(t)}\right)^2 \frac{w(t)}{f(t)-f(z)}\frac{dt}{t}.$$

Definition 2.2. For $n \in \mathbb{Z}$, let

$$L_n \stackrel{\text{def}}{=} -iK_{e_n},$$

a differential operator in the variables c_n.

For nonnegative n, it is very easy to compute L_n.

Proposition 2.1.

(1) *For* $n \geq 1$, $L_n = \dfrac{\partial}{\partial c_n} + \displaystyle\sum_{k=1}^{+\infty}(k+1)c_k\frac{\partial}{\partial c_{n+k}}$;

(2) $L_0 = \displaystyle\sum_{n\geq 1} nc_n\frac{\partial}{\partial c_n}$.

Proof.

(1) In this case, the expression for K_v becomes, with $t = e^{i\theta}$,

$$\begin{aligned}
K_{e_n}(f)(z) &= \frac{f(z)^2}{2\pi}\int_0^{2\pi}\left(\frac{tf'(t)}{f(t)}\right)^2 \frac{t^n}{f(t)-f(z)}\frac{dt}{t}\\
&= \frac{f(z)^2}{2\pi}\int_0^{2\pi}\left(\frac{tf'(t)}{f(t)}\right)^2 \frac{t^{n-1}}{f(t)-f(z)}dt\\
&= \frac{f(z)^2}{2\pi}2i\pi\operatorname{Res}_z\left[\left(\frac{tf'(t)}{f(t)}\right)^2 \frac{t^{n-1}}{f(t)-f(z)}\right]\\
&\qquad\qquad\text{(by Cauchy's formula)}\\
&= \frac{f(z)^2}{2\pi}2i\pi\left(\frac{zf'(z)}{f(z)}\right)^2 \frac{z^{n-1}}{f'(z)}\\
&= iz^{n+1}f'(z)\\
&= iz^{n+1} + i\sum_{k=1}^{+\infty}(k+1)c_k z^{k+n+1},
\end{aligned}$$

therefore

$$L_n(f)(z) = z^{n+1} + \sum_{k=1}^{+\infty}(k+1)c_k z^{k+n+1},$$

whence the result.

(2) The computation is similar, taking into account the pole at 0, and yields

$$L_0(f)(z) = zf'(z) - f(z),$$

whence the result.

□

Lemma 2.1. *One has the commutation relations, for any $(m, n) \in \mathbb{Z}^2$,*

$$[L_m, L_n] = (m-n)L_{m+n}\,. \qquad (*)$$

Proof. [2, p. 655] for $m \geq 0$ and $n \geq 0$, it is actually very easy to check this relation using Proposition 2.1. □

3. The Neretin polynomials and the representation ρ

Let $\gamma_k \stackrel{\text{def}}{=} \frac{c}{12}(k^3 - k)$, and $P_k = 0$ for $k < 0$.

Theorem 3.1 (Kirillov-Neretin). *There exists a unique sequence $(P_n)_{n\geq 0}$ of polynomials in the $(c_i)_{i\geq 1}$ such that:*

(1) P_k depends only upon $c_1,\ldots,c_k$;
(2) $P_0 = h$;
(3) $\forall k \geq 1\, \forall n \geq 1\ L_k(P_n) = (n+k)P_{n-k} + \gamma_k \delta_{k,n}$;
(4) $\forall n \geq 1\ P_n(0) = 0$.

Proof. Given $P_0, \ldots, P_n$ $(n \geq 0)$, the relations (3) (with $n+1$ in place of n) are trivially satisfied for any polynomial P_{n+1} in $c_1, \ldots, c_{n+1}$ and any $k > n+1$; for $1 \leq k \leq n+1$, the relations determine, by descending induction on k, the $\frac{\partial P_{n+1}}{\partial c_k}$ in a unique way, therefore they determine P_{n+1} up to a constant; (4) for $n+1$ now determines a unique P_{n+1}. □

The first few terms of the sequence are easily computed:

$$P_0 = h\,,$$
$$P_1 = 2hc_1\,,$$
$$P_2 = \left(4h + \frac{c}{2}\right)c_2 - \left(h + \frac{c}{2}\right)c_1^2\,.$$

If each c_k is given the weight k, it is easily seen that P_k is homogeneous of weight k.

Let us remind the reader of the definition of the *Schwarzian derivative* of a holomorphic function f:

$$S(f)(z) \stackrel{\text{def}}{=} \frac{f'''(z)}{f'(z)} - \frac{3}{2}\left(\frac{f''(z)}{f'(z)}\right)^2 .$$

The following result could have been used as definition of the polynomials P_k:

Proposition 3.1 ([9, p. 742, Theorem]). *For any $f \in \mathcal{M}$*

$$\sum_{n=0}^{+\infty} P_n(c_1, \ldots, c_n) z^n = h\left(\frac{zf'(z)}{f(z)}\right)^2 + \frac{cz^2}{12} S(f)(z) .$$

Proposition 3.2. *For any $k \geq 0$, $p \geq 0$,*

$$L_{-k}(P_p) - L_{-p}(P_k) = (p-k)P_{p+k} ;$$

in particular, the formula of Theorem 3.1(3) remains valid for $k = 0$.

Proof. [2, p. 663]. □

Let

$$Q_k \stackrel{\text{def}}{=} \begin{cases} P_k & \text{for } k \neq 0 , \\ 0 & \text{for } k = 0 . \end{cases}$$

Theorem 3.2. *Let us set, for each $k \in \mathbb{Z}$,*

$$\rho(e_k) = -i\,(L_k + Q_{-k})$$

and

$$\rho(\kappa) = i\,\mathrm{Id} .$$

Then ρ defines a representation of the Lie algebra $\mathcal{V}ir_{c,h}$ into the Lie algebra of differential operators on $\mathcal{M}$.

Proof. As, obviously, $[\rho(e_k), \rho(\kappa)] = 0$ is enough to prove that

$$[\rho(e_m), \rho(e_n)] = \rho([e_m, e_n]) .$$

Taking Proposition 1.1 into account, this is easily reduced to checking the relation:

$$L_n(Q_{-m}) - L_m(Q_{-n}) = (n-m)Q_{-m-n} - [2hm + \gamma_m]\delta_{m,-n} .$$

But, for $m \geq 0$ and $n \geq 0$, that relation is trivially satisfied; for $m = 0$ and $n < 0$, as well as for $n = 0$ and $m < 0$, it follows from the relation for any $n \geq 1$

$$L_0(P_n) = nP_n\,;$$

in the case $m < 0$ and $n < 0$, setting $p = -m$ and $k = -n$, it is enough to prove that for any $p \geq 1$ and $k \geq 1$

$$L_{-k}(P_p) - L_{-p}(P_k) = (p-k)P_{p+k}\,,$$

but both these facts follow from Proposition 3.2.

There remains the case $m \leq -1$ and $n \geq 1$ (or the other way round); in this case, we need to prove, setting $k = -m \geq 1$, that:

$$\forall k \geq 1\, \forall n \geq 1\; L_n(P_k) = (n+k)Q_{k-n} + (2hk + \gamma_k)\delta_{k,n}$$

i.e., for any $k \geq 1$ and $n \geq 1$,

$$L_n(P_k) = \begin{cases} (n+k)P_{k-n} & \text{if } n \neq k\,, \\ 2hk + \gamma_k & \text{if } n = k\,. \end{cases}$$

As $P_0 = h$, this follows from Theorem 3.1(3). □

4. Definition of an unitarizing measure and a non-existence result

Definition 4.1. A Borel probability measure μ on $\mathcal{M}$ is said to be unitarizing for the representation ρ if and only if for any $v \in \mathcal{V}ir^{\mathbb{R}}_{c,h}$

$$\rho(v)^* = -\rho(v)$$

on the space $\mathcal{H}L^2_\mu(\mathcal{M})$ of μ-square integrable holomorphic functions on $\mathcal{M}$.

Lemma 4.1 ([1, Theorem 1, p. 433]). *If μ exists, then, setting $Z_k = L_k - \overline{L_{-k}} (k \geq 0)$, one has for any $F \in \mathcal{C}^\infty(\mathcal{M})$*

$$\int_{\mathcal{M}} Z_k(F)d\mu = -\int_{\mathcal{M}} F\beta_k d\mu\,, \tag{1}$$

where

$$\beta_k = \begin{cases} -\bar{P}_k & \text{if } k \geq 1\,, \\ 0 & \text{if } k = 0\,. \end{cases}$$

Proof. From the definition follows that for any $v \in \mathcal{V}ir_{c,h}$

$$\rho(v)^* = -\rho(\bar{v}) .$$

By a density argument, one may assume that $F = \varphi\bar{\psi}$, with φ and ψ holomorphic; then

$$\begin{aligned}
\int_{\mathcal{M}} Z_k(F)d\mu &= \int_{\mathcal{M}} L_k(\varphi\bar{\psi})d\mu - \int_{\mathcal{M}} \overline{L_{-k}}(\varphi\bar{\psi})d\mu \\
&= \int_{\mathcal{M}} L_k(\varphi\bar{\psi})d\mu - \overline{\int_{\mathcal{M}} L_{-k}(\bar{\varphi}\psi)d\mu} \\
&= \int_{\mathcal{M}} (L_k(\varphi)\bar{\psi} + \varphi L_k(\bar{\psi}))d\mu - \overline{\int_{\mathcal{M}} (L_{-k}(\bar{\varphi})\psi + \bar{\varphi}L_{-k}(\psi))d\mu} \\
&= \int_{\mathcal{M}} (L_k(\varphi)\bar{\psi} + \varphi L_k(\bar{\psi}))d\mu - \overline{\int_{\mathcal{M}} (L_{-k}(\bar{\varphi})\psi + \bar{\varphi}L_{-k}(\psi))d\mu} \\
&= \int_{\mathcal{M}} L_k(\varphi)\bar{\psi}d\mu - \overline{\int_{\mathcal{M}} \bar{\varphi}L_{-k}(\psi)d\mu} \\
&= \int_{\mathcal{M}} (\rho(e_k)(\varphi) - Q_{-k}\varphi)\bar{\psi}d\mu - \overline{\int_{\mathcal{M}} \bar{\varphi}(i\rho(e_{-k})(\psi) - Q_k\psi)d\mu} \\
&= i(\rho(e_k)(\varphi), \psi) + \int_{\mathcal{M}} \varphi(\bar{Q_k} - Q_{-k})\bar{\psi}d\mu + i\overline{(\rho(\overline{e_k})(\psi), \varphi)} \\
&= i(\rho(e_k)(\varphi), \psi) + \int_{\mathcal{M}} \varphi(\bar{Q_k} - Q_{-k})\bar{\psi}d\mu + i(\varphi, \rho(\bar{e_k})(\psi)) \\
&= \int_{\mathcal{M}} \varphi(\bar{Q_k} - Q_{-k})Fd\mu ,
\end{aligned}$$

by the hypothesis on μ and since φ and ψ are holomorphic.

Whence we have the result with

$$\beta_k = Q_{-k} - \overline{Q_k} = \begin{cases} -\overline{P_k} & \text{for } k \geq 1 , \\ 0 & \text{for } k = 0 . \end{cases}$$

□

Theorem 4.1 ([1, Theorem 3 and Corollary 4, p. 234]).

(1) *If μ exists then the sequence $1, P_1, P_2, dots$ is a sequence of orthogonal polynomials in $L^2(\mathcal{M}, \mu)$; more precisely:*

$$(P_m, P_k)_{L^2(\mu)} = \begin{cases} 0 & \textit{if } m \neq k , \\ \gamma_k + 2hk & \textit{if } m = k \geq 1 , \\ h^2 & \textit{if } m = k = 0 . \end{cases}$$

(2) *If $h = 0$ then there is no unitarizing measure on $\mathcal{M}$ for ρ.*

Proof.

(1) Let us set, for each $k \geq 0$, and $H_k = Z_k^2 + \beta_k Z_k$; it follows from Lemma 4.1 applied to $Z_k(F)$ that, for each $k \geq 0$, one has: for any $F \in \mathcal{C}^\infty(\mathcal{M})$ and $k \geq 0$

$$\int_{\mathcal{M}} H_k(F)\, d\mu = 0\,. \tag{2}$$

But it follows from the definition of the Neretin polynomials (Theorem 3.1(3)) and from the last remark in Proposition 3.2 that for any $k \geq 0$ and $n \geq 1$

$$\begin{aligned} H_k(P_n) &= L_k((n+k)P_{n-k} + \gamma_k \delta_{k,n}) + \beta_k((n+k)P_{n-k} + \gamma_k \delta_{k,n}) \\ &= (n+k)nP_{n-2k} + (n+k)\gamma_k\delta_{k,n-k} + (n+k)\beta_k P_{n-k} \\ &\quad + \beta_k\gamma_k\delta_{k,n}\,. \end{aligned} \tag{3}$$

By (2) one has for any $k \geq 0$ and $n \geq 1$

$$\int_{\mathcal{M}} H_k(P_n)\, d\mu = 0\,. \tag{4}$$

Applying (3) for $k = 0$ and $n \geq 1$, one finds that for any $n \geq 1$

$$H_0(P_n) = n^2 P_n\,,$$

whence (4) yields that for any $n \geq 1$

$$\int_{\mathcal{M}} P_n\, d\mu = 0\,. \tag{5}$$

From Lemma 4.1 applied to $F = 1$ follows that for any $k \geq 0$

$$\int_{\mathcal{M}} \beta_k d\mu = 0\,. \tag{6}$$

Taking now $k \geq 1$, $m \geq 1$ and $n = m+k$, (3) and (4) together yield:

$$\begin{aligned} \int_{\mathcal{M}} \big[(2k+m)(k+m)P_{m-k} &+ (m+2k)\gamma_k\delta_{m,k} \\ &+ (m+2k)\beta_k P_m + \beta_k\gamma_k\delta_{m,0}\big]\, d\mu = 0\,; \end{aligned}$$

from the fact that

$$\int_{\mathcal{M}} P_n \, d\mu = \begin{cases} 0 & \text{for } n \geq 1 \, (5) \\ h & \text{for } n = 0 \\ 0 & \text{for } n < 0 \, (\text{by definition}) \end{cases}$$

and from (6), we get:

$$\int_{\mathcal{M}} \beta_k P_m \, d\mu = \begin{cases} 0 & \text{if } m \neq k \\ -\gamma_k - 2kh & \text{if } m = k \neq 0 \, . \end{cases}$$

Recall that $\beta_k = -\bar{P}_k$ for $k \geq 1$ and therefore the result holds.

(2) Let us remind the reader that $P_1 = 2hc_1$. Clearly,

$$Z_1(c_1) = (L_1 - \overline{L_{-1}})(c_1) = L_1(c_1) = 1 \, ,$$

whence

$$\begin{aligned} 1 &= \int_{\mathcal{M}} d\mu = \int_{\mathcal{M}} Z_1(c_1) d\mu = -\int_{\mathcal{M}} c_1 \beta_1 \, d\mu \text{ (by Lemma 4.1)} \\ &= \int_{\mathcal{M}} c_1 \bar{P}_1 \, d\mu = 2h \int_{\mathcal{M}} c_1 \bar{c}_1 \, d\mu \, , \end{aligned}$$

which is impossible for $h = 0$. A more geometrical proof of this nonexistence result had previously been given in [3, Theorem 2.2, p. 625].

□

References

1. H. Airault, V. Bogachev, "Realization of Virasoro unitarizing measures on the set of Jordan curves", *C.R.Acad.Sci.Paris, Ser. I* **336**, 429–434 (2003).
2. H. Airault, P. Malliavin, "Unitarizing probability measures for representations of Virasoro algebra", *J. Math. Pures Appl.* **80**(6), 627–667 (2001).
3. H. Airault, P. Malliavin, A. Thalmaier, "Support of Virasoro unitarizing measures", *C. R. Acad. Sci. Paris, Ser. I* **335**, 621–626 (2002).
4. H. Airault, P. Malliavin, A. Thalmaier, "Canonical Brownian motion on the space of univalent functions and resolution of Beltrami equations by a continuity method along stochastic flows", *J. Math. Pures Appl.* **83**, 955–1018 (2004).
5. M. Gordina, P. Lescot, "Riemannian Geometry of Diff$(S^1)/S^1$", *Journal of Functional Analysis* (2006).

6. V. G. Kac, *Infinite Dimensional Lie Algebras.*
7. A. A. Kirillov, "Kähler structures on K-orbits of the group of diffeomorphisms of a circle".
8. A. A. Kirillov, D. V. Yuriev, "Representations of the Virasoro algebra by the orbit method", *J. Geom. Phys.* **5**, no. 3, 351–363 (1988).
9. A. A. Kirillov, "Geometric approach to discrete series of unirreps for VIR", *J. Math. Pures Appl.* **77**, 735–746 (1998).

Paul Lescot
INSSET – Université de Picardie,
48 Rue Raspail, 02100 Saint-Quentin, France
paul.lescot@u-picardie.fr

FKG INEQUALITY ON THE WIENER SPACE VIA PREDICTABLE REPRESENTATION

Yutao Ma (*Wuhan University, Hubei / Université de La Rochelle*),
Nicolas Privault (*Université de La Rochelle*)

Using the Clark predictable representation formula, we give a proof of the FKG inequality on the Wiener space. Solutions of stochastic differential equations are treated as applications and we recover by a simple argument the covariance inequalities obtained for diffusions processes by several authors.

1. Introduction

Let $(\Omega, \mathcal{F}, \mathbb{P}, \preceq)$ be a probability space equipped with a partial order relation $\preceq$ on Ω. An (everywhere defined) real-valued random variable F on $(\Omega, \mathcal{F}, \mathbb{P}, \preceq)$ is said to be non-decreasing if

$$F(\omega_1) \leq F(\omega_2)$$

for any $\omega_1, \omega_2 \in \Omega$ satisfying $\omega_1 \preceq \omega_2$. The FKG inequality [4] states that if F and G are two square-integrable random functionals which are non-decreasing for the order $\preceq$, then F and G are non-negatively correlated:

$$\mathrm{Cov}(F, G) \geq 0.$$

It is well known that the FKG inequality holds for the standard ordering on $\Omega = \mathbb{R}$, since given $X, Y : \mathbb{R} \to \mathbb{R}$ two non-decreasing functions on $\mathbb{R}$ we have:

$$\begin{aligned}
\mathrm{Cov}(X, Y) &= \frac{1}{2} \int_{\mathbb{R}\times\mathbb{R}} (X(x) - X(y))\,(Y(x) - Y(y))\,\mathbb{P}(dx)\,\mathbb{P}(dy) \\
&= \frac{1}{2} \int_{\{y \leq x\}} (X(x) - X(y))\,(Y(x) - Y(y))\,\mathbb{P}(dx)\,\mathbb{P}(dy) \\
&+ \frac{1}{2} \int_{\{x < y\}} (X(y) - X(x))\,(Y(y) - Y(x))\,\mathbb{P}(dx)\,\mathbb{P}(dy) \\
&\geq 0.
\end{aligned}$$

The FKG inequality also holds on $\mathbb{R}^n$ for the pointwise ordering, cf. e.g. Bakry and Michel [2].

On the Wiener space $(\Omega, \mathcal{F}, \mathbb{P})$ with Brownian motion $(W_t)_{t\in\mathbb{R}_+}$, Barbato [3] introduced a weak ordering on continuous functions and proved an FKG inequality for Wiener functionals, with application to diffusion processes.

In this paper we recover the results of [3] under weaker hypotheses via a simple argument. Our approach is inspired by Remark 1.5 stated on the Poisson space in Wu [14], page 432, which can be carried over to the Wiener space by saying that the *predictable representation* of a random variable F as an Itô integral, obtained via the Clark formula

$$F = \mathbb{E}[F] + \int_0^1 \mathbb{E}[D_t F|\mathcal{F}_t]\, dW_t,$$

yields the covariance identity

$$\mathrm{Cov}(F,G) = \mathbb{E}\left[\int_0^1 \mathbb{E}[D_t F|\mathcal{F}_t]\,\mathbb{E}[D_t G|\mathcal{F}_t]\, dt\right], \tag{1}$$

where D is the Malliavin gradient expressed as

$$\langle DF, \dot{h}\rangle_{L^2([0,1])} = \frac{d}{d\varepsilon} F(\omega + \epsilon h)\Big|_{\varepsilon=0}. \tag{2}$$

From (2) we deduce that DF is non-negative when F is non-decreasing, which implies $\mathrm{Cov}(F,G) \geq 0$ from (1). Applications are given to diffusion processes and in Theorem 3.2 we recover, under weaker hypotheses, the covariance inequality obtained in Theorem 3.2 of [7] and in Theorem 7 of [3].

We proceed as follows. Elements of analysis on the Wiener space and applications to covariance identities are recalled in Section 2. The FKG inequality and covariance inequalities for diffusions are proved in Section 3. We also show that our method allows us to deal with the discrete case, cf. Section 4.

2. Analysis on the Wiener space

In this section we recall some elements of stochastic analysis on the classical Wiener space $(\Omega, \mathcal{F}, \mathbb{P})$ on $\Omega = C_0([0,1];\mathbb{R})$, with canonical Brownian motion $(W_t)_{t\in[0,1]}$ generating the filtration $(\mathcal{F}_t)_{t\in[0,1]}$. Our results extend without difficulty to the Wiener space on $C_0(\mathbb{R}_+;\mathbb{R})$. Let H denote the Cameron–Martin space, i.e., the space of absolutely continuous functions with square-integrable derivative:

$$H = \left\{ h : [0,1] \to \mathbb{R} \, : \, \int_0^1 |\dot{h}(s)|^2\, ds < \infty \right\}.$$

Let $I_n(f_n)$, $n \geq 1$, denote the iterated stochastic integral of f_n in the space $L_s^2([0,1]^n)$ of symmetric square-integrable functions in n variables on $[0,1]^n$, defined as

$$I_n(f_n) = n! \int_0^1 \int_0^{t_n} \cdots \int_0^{t_2} f_n(t_1, \ldots, t_n)\, dW_{t_1} \cdots dW_{t_n},$$

with the isometry formula

$$\mathbb{E}[I_n(f_n)\, I_m(g_m)] = n!\, 1_{\{n=m\}} \langle f_n, g_m \rangle_{L^2([0,1]^n)}.$$

Every $F \in L^2(\Omega)$ admits a unique Wiener chaos expansion

$$F = \mathbb{E}[F] + \sum_{n=1}^{\infty} I_n(f_n)$$

with $f_n \in L_s^2([0,1]^n)$, $n \geq 1$. Let $(e_k)_{k \geq 1}$ denote the dyadic basis of $L^2([0,1])$ given by

$$e_k = 2^{n/2}\, \mathbf{1}_{\left[\frac{k-2^n}{2^n}, \frac{k+1-2^n}{2^n}\right]}, \qquad 2^n \leq k \leq 2^{n+1} - 1, \quad n \in \mathbb{N}.$$

Recall the following two equivalent definitions of the Malliavin gradient D and its domain $\mathrm{Dom}(D)$, cf. Lemma 1.2 of [8] and [10]:

a) *Finite dimensional approximations.* Given $F \in L^2(\Omega)$, let for all $n \in \mathbb{N}$

$$\mathcal{G}_n = \sigma(I_1(e_{2^n}), \ldots, I_1(e_{2^{n+1}-1})),$$

and $F_n = \mathbb{E}[F|\mathcal{G}_n]$, and consider f_n a square-integrable function with respect to the standard Gaussian measure on $\mathbb{R}^{2^n}$, such that

$$F_n = f_n(I_1(e_{2^n}), \ldots, I_1(e_{2^{n+1}-1})).$$

Then $F \in \mathrm{Dom}(D)$ if and only if f_n belongs for all $n \geq 1$ to the Sobolev space $W^{2,1}(\mathbb{R}^{2^n})$ with respect to the standard Gaussian measure on $\mathbb{R}^{2^n}$, and the sequence

$$D_t F_n := \sum_{i=1}^{2^n} e_{2^n+i-1}(t) \frac{\partial f_n}{\partial x_i}(I_1(e_{2^n}), \ldots, I_1(e_{2^{n+1}-1})), \qquad t \in [0,1],$$

converges in $L^2(\Omega \times [0,1])$. In this case we let

$$DF := \lim_{n \to \infty} DF_n.$$

b) *Chaos expansions.* Let $G \in L^2(\Omega)$ be given by

$$G = \mathbb{E}[G] + \sum_{n=1}^{\infty} I_n(g_n).$$

Then G belongs to $\mathrm{Dom}(D)$ if and only if the series

$$\sum_{n=1}^{\infty} n!\, n\, \|g_n\|^2_{L^2([0,1]^n)}$$

converges and, in this case,

$$D_t G = g_1(t) + \sum_{n=1}^{\infty} n\, I_{n-1}(g_n(*, t)), \qquad t \in [0,1].$$

In case (a) above the gradient $\langle DF_n, \dot{h}\rangle_{L^2([0,1])}$, $h \in H$, coincides with the directional derivative

$$\begin{aligned}\langle DF_n, \dot{h}\rangle_{L^2([0,1])} &= \frac{d}{d\varepsilon} f_n\big(I_1(e_{2^n}) + \varepsilon\langle e_{2^n}, \dot{h}\rangle_{L^2([0,1])}, \ldots, \\ &\qquad I_1(e_{2^{n+1}-1}) + \varepsilon\langle e_{2^{n+1}-1}, \dot{h}\rangle_{L^2([0,1])}\big)\Big|_{\varepsilon=0} \\ &= \frac{d}{d\varepsilon} F_n(\omega + \epsilon h)\Big|_{\varepsilon=0},\end{aligned}$$

where the limit exists in $L^2(\Omega)$.

Similarly, the Ornstein–Uhlenbeck semi-group $(P_t)_{t\in\mathbb{R}_+}$ admits the following equivalent definitions, cf. e.g. [9], [12], [13]:

a) *Integral representation.* For any $F \in L^2(\Omega)$ and $t \in \mathbb{R}_+$, let

$$P_t F(\omega) = \int_\Omega F\big(e^{-t}\omega + \sqrt{1 - e^{-2t}}\,\tilde{\omega}\big)\, d\mathbb{P}(\tilde{\omega}), \qquad \mathbb{P}(d\omega)\text{-a.s.} \quad (3)$$

b) *Chaos representation.* For any $F \in L^2(\Omega)$ with the chaos expansion

$$F = \mathbb{E}[F] + \sum_{n=1}^{\infty} I_n(f_n),$$

we have

$$P_t F = \mathbb{E}[F] + \sum_{n=1}^{\infty} e^{-nt} I_n(f_n), \qquad t \in \mathbb{R}_+. \quad (4)$$

The operator D satisfies the Clark formula, i.e.,

$$F = \mathbb{E}[F] + \int_0^1 \mathbb{E}[D_t F|\mathcal{F}_t]\, dW_t, \qquad F \in \mathrm{Dom}(D), \tag{5}$$

cf. e.g. [12]. By continuity of the operator mapping $F \in L^2(\Omega)$ to the adapted and square-integrable process $(u_t)_{t\in\mathbb{R}_+}$ appearing in predictable representation

$$F = \mathbb{E}[F] + \int_0^1 u_t\, dW_t, \tag{6}$$

the Clark formula can be extended to any $F \in L^2(\Omega)$ as in the following proposition.

Proposition 2.1. *The operator $F \mapsto (\mathbb{E}[D_t F|\mathcal{F}_t])_{t\in[0,1]}$ extends as a continuous operator on $L^2(\Omega)$.*

Proof. We use the bound

$$\mathbb{E}\left[\int_0^1 (\mathbb{E}[D_t F|\mathcal{F}_t])^2\, dt\right] = \mathbb{E}[(F-\mathbb{E}[F])^2] = \mathbb{E}[F^2] - (\mathbb{E}[F])^2 \le \|F\|^2_{L^2(\Omega)}, \tag{7}$$

for $F \in \mathrm{Dom}(D)$. □

Moreover, by uniqueness of the predictable representation of $F \in L^2(\Omega)$, an expression of the form

$$F = c + \int_0^1 u_t\, dW_t,$$

where $c \in \mathbb{R}$ and $(u_t)_{t\in\mathbb{R}_+}$ is adapted and square-integrable, implies $u_t = \mathbb{E}[D_t F|\mathcal{F}_t]$, $dt \times d\mathbb{P}$-a.e. The Clark formula and the Itô isometry yield the following covariance identity, cf. Proposition 2.1 of [6].

Proposition 2.2. *For any $F, G \in L^2(\Omega)$ we have*

$$\mathrm{Cov}(F,G) = \mathbb{E}\left[\int_0^1 \mathbb{E}[D_t F|\mathcal{F}_t]\, \mathbb{E}[D_t G|\mathcal{F}_t]\, dt\right]. \tag{8}$$

This identity can be written as

$$\mathrm{Cov}(F,G) = \mathbb{E}\left[\int_0^1 \mathbb{E}[D_t F|\mathcal{F}_t]\, D_t G\, dt\right], \tag{9}$$

provided $G \in \mathrm{Dom}(D)$. The following lemma is an immediate consequence of (8).

Lemma 2.1. *Let $F, G \in L^2(\Omega)$ such that*

$$\mathbb{E}[D_t F|\mathcal{F}_t] \cdot \mathbb{E}[D_t G|\mathcal{F}_t] \geq 0, \qquad dt \times d\mathbb{P}\text{-a.e.}$$

Then F and G are non-negatively correlated:

$$\mathrm{Cov}(F, G) \geq 0.$$

If $G \in \mathrm{Dom}(D)$, resp. $F, G \in \mathrm{Dom}(D)$, the above condition can be replaced by

$$\mathbb{E}[D_t F|\mathcal{F}_t] \geq 0 \quad \text{and} \quad D_t G \geq 0, \qquad dt \times d\mathbb{P}\text{-a.e.},$$

resp.

$$D_t F \geq 0 \quad \text{and} \quad D_t G \geq 0, \qquad dt \times d\mathbb{P}\text{-a.e.}.$$

As recalled in the introduction, if X is a real random variable and f, g are $\mathcal{C}^1(\mathbb{R})$ functions with non-negative derivatives f', g', then $f(X)$ and $g(X)$ are non-negatively correlated. Lemma 2.1 provides an analog of this result on the Wiener space, replacing the ordinary derivative with the adapted process $(\mathbb{E}[D_t F|\mathcal{F}_t])_{t\in[0,1]}$.

3. FKG inequality on the Wiener space

We consider the order relation introduced in [3].

Definition 3.1. Given $\omega_1, \omega_2 \in \Omega$, we say that $\omega_1 \preceq \omega_2$ if and only if we have

$$\omega_1(t_2) - \omega_1(t_1) \leq \omega_2(t_2) - \omega_2(t_1), \qquad 0 \leq t_1 \leq t_2 \leq 1.$$

The class of non-decreasing functionals with respect to $\preceq$ is larger than that of non-decreasing functionals with respect to the pointwise order on Ω defined by

$$\omega_1(t) \leq \omega_2(t), \qquad t \in [0,1], \quad \omega_1, \omega_2 \in \Omega.$$

Definition 3.2. A random variable $F : \Omega \to \mathbb{R}$ is said to be non-decreasing if

$$\omega_1 \preceq \omega_2 \Rightarrow F(\omega_1) \leq F(\omega_2), \qquad \mathbb{P}(d\omega_1) \otimes \mathbb{P}(d\omega_2)\text{-a.s.}$$

Note that unlike in [3], the above definition allows for almost-surely defined functionals. The next result is the FKG inequality on the Wiener space. It recovers Theorem 4 of [3] under weaker (i.e., almost-sure) hypotheses.

Theorem 3.1. *For any non-decreasing functionals $F, G \in L^2(\Omega)$ we have*

$$\mathrm{Cov}(F, G) \geq 0.$$

The proof of this result is a direct consequence of Lemma 2.1 and Proposition 3.1 below.

Lemma 3.1. *For every non-decreasing $F \in \mathrm{Dom}(D)$ we have*

$$D_t F \geq 0, \qquad dt \times d\mathbb{P}\text{-a.e.}.$$

Proof. For $n \in \mathbb{N}$, let π_n denote the orthogonal projection from $L^2([0,1])$ onto the linear space generated by $(e_k)_{2^n \leq k < 2^{n+1}}$. Consider h in the Cameron–Martin space H and let

$$h_n(t) = \int_0^t [\pi_n \dot{h}](s)\, ds, \qquad t \in [0,1], \quad n \in \mathbb{N}.$$

Let Λ_n denote the square-integrable and $\mathcal{G}_n$-measurable random variable

$$\Lambda_n = \exp\left(\int_0^1 [\pi_n \dot{h}](s)\, dW_s - \frac{1}{2} \int_0^1 |[\pi_n \dot{h}](s)|^2\, ds \right).$$

From the Cameron–Martin theorem, for all $n \in \mathbb{N}$ and $\mathcal{G}_n$-measurable bounded random variable G_n we have, letting $F_n = \mathbb{E}[F \mid \mathcal{G}_n]$:

$$\begin{aligned}
\mathbb{E}[F_n(\cdot + h_n) G_n] &= \mathbb{E}[\Lambda_n\, F_n\, G_n(\cdot - h_n)] \\
&= \mathbb{E}[\Lambda_n\, \mathbb{E}[F|\mathcal{G}_n]\, G_n(\cdot - h_n)] \\
&= \mathbb{E}[\mathbb{E}[\Lambda_n\, F\, G_n(\cdot - h_n) \mid \mathcal{G}_n]] \\
&= \mathbb{E}[\Lambda_n\, F\, G_n(\cdot - h_n)] \\
&= \mathbb{E}[F(\cdot + h_n)\, G_n],
\end{aligned}$$

hence

$$F_n(\omega + h_n) = \mathbb{E}[F(\cdot + h_n) | \mathcal{G}_n](\omega), \qquad \mathbb{P}(d\omega)\text{-a.s.}$$

If $\dot{h}$ is non-negative, then $\pi_n \dot{h}$ is non-negative by construction hence $\omega \preceq \omega + h_n$, $\omega \in \Omega$, and we have

$$F(\omega) \leq F(\omega + h_n), \qquad \mathbb{P}(d\omega)\text{-a.s.},$$

since, from the Cameron–Martin theorem, $\mathbb{P}(\{\omega + h_n \,:\, \omega \in \Omega\}) = 1$. Hence with the notation of Section 2,

$$\begin{aligned}
F_n(\omega + h) &= f_n\big(I_1(e_{2^n}) + \langle e_{2^n}, \dot{h}\rangle_{L^2([0,1])}, \ldots, \\
&\qquad I_1(e_{2^{n+1}-1}) + \langle e_{2^{n+1}-1}, \dot{h}\rangle_{L^2([0,1])}\big) \\
&= f_n\big(I_1(e_{2^n}) + \langle e_{2^n}, \pi_n\dot{h}\rangle_{L^2([0,1])}, \ldots, \\
&\qquad I_1(e_{2^{n+1}-1}) + \langle e_{2^{n+1}-1}, \pi_n\dot{h}\rangle_{L^2([0,1])}\big) \\
&= F_n(\omega + h_n) \\
&= \mathbb{E}[F(\cdot + h_n)|\mathcal{G}_n](\omega) \\
&\geq \mathbb{E}[F|\mathcal{G}_n](\omega) \\
&= F_n(\omega), \qquad \mathbb{P}(d\omega)\text{-a.s.},
\end{aligned}$$

i.e., for any $\varepsilon_1 \leq \varepsilon_2$ and $h \in H$ such that $\dot{h}$ is non-negative we have

$$F_n(\omega + \varepsilon_1 h) \leq F_n(\omega + \varepsilon_2 h),$$

and the smooth function $\varepsilon \mapsto F_n(\omega + \varepsilon h)$ is non-decreasing in ε on $\mathbb{R}$, $\mathbb{P}(d\omega)$-a.s. As a consequence,

$$\langle DF_n, \dot{h}\rangle_{L^2([0,1])} = \frac{d}{d\varepsilon} F_n(\omega + \epsilon h)\Big|_{\varepsilon=0} \geq 0,$$

for all $h \in H$ such that $\dot{h} \geq 0$, hence $DF_n \geq 0$. Taking the limit of $(DF_n)_{n\in\mathbb{N}}$ as n goes to infinity shows that $DF \geq 0$. □

Next, we extend Lemma 3.1 to $F \in L^2(\Omega)$.

Proposition 3.1. *For any non-decreasing functional $F \in L^2(\Omega)$ we have*

$$\mathbb{E}[D_t F \mid \mathcal{F}_t] \geq 0, \qquad dt \times d\mathbb{P}\text{-}a.e.$$

Proof. Assume that $F \in L^2(\Omega)$ is non-decreasing. Then $P_{1/n}F$, $n \geq 1$, is non-decreasing from (3), and belongs to $\mathrm{Dom}(D)$ from (4). From Lemma 3.1 we have

$$D_t P_{1/n} F \geq 0, \qquad dt \times d\mathbb{P}\text{-a.e.},$$

hence

$$\mathbb{E}[D_t P_{1/n} F \mid \mathcal{F}_t] \geq 0, \qquad dt \times d\mathbb{P}\text{-a.e.}$$

Taking the limit as n goes to infinity yields $\mathbb{E}[D_t F \mid \mathcal{F}_t] \geq 0$, $dt \times d\mathbb{P}$-a.e. from (7) and the fact that $P_{1/n}F$ converges to F in $L^2(\Omega)$ as n goes to infinity. □

Conversely it is not difficult to show that if $u \in L^2([0,1])$ is a non-negative deterministic function, then the Wiener integral $\int_0^1 u_t dW_t$ is a non-decreasing functional. Note however that the stochastic integral of a non-negative square-integrable process may not necessarily be a non-decreasing functional. For example, consider $u_t = G1_{[a,1]}$, $t \in [0,1]$, where $G \in L^2(\Omega, \mathcal{F}_a)$ is non-negative and decreasing, then

$$\int_0^1 u_t \, dW_t = G(B_1 - B_a)$$

is not non-decreasing.

Example: maximum of Brownian motion
By Proposition 2.1.3 of [9] the maximum $M = \sup_{0 \le t \le 1} W(t)$ of Brownian motion on $[0,1]$ belongs to $\mathrm{Dom}(D)$ and satisfies $DM = 1_{[0,\tau]} \ge 0$, where τ is the a.s. unique point where M attains its maximum. Here, M is clearly an increasing functional.

Example: diffusion processes
Consider the stochastic differential equations

$$\begin{cases} dX_t = b_t(X_t)\, dt + \sigma_t(X_t)\, dW_t, \\ X_0 = x_0, \end{cases} \tag{10}$$

and

$$\begin{cases} d\tilde{X}_t = \tilde{b}_t(\tilde{X}_t)\, dt + \tilde{\sigma}_t(\tilde{X}_t)\, dW_t, \\ \tilde{X}_0 = x_0, \end{cases} \tag{11}$$

where b, $\tilde{b}$, σ, $\tilde{\sigma}$ are functions on $\mathbb{R}_+ \times \mathbb{R}$ satisfying the following global Lipschitz and boundedness conditions, cf. [9], page 99:

(i) $|\sigma_t(x) - \sigma_t(y)| + |b_t(x) - b_t(y)| \le K|x-y|, \quad x, y \in \mathbb{R},\ t \in [0,1],$
(ii) $t \mapsto \sigma_t(0)$ and $t \mapsto b_t(0)$ are bounded on $[0,1]$,

for some $K > 0$. Lemma 8 of [3] shows that the solutions $(X_t)_{t\in[0,1]}$, $(\tilde{X}_t)_{t\in[0,1]}$ of (10) and (11) are increasing functionals when $\sigma(x)$, $\tilde{\sigma}(x)$ are differentiable with Lipschitz derivative in one variable and satisfy uniform bounds of the form

$$0 < \varepsilon \le \sigma(x) \le M < \infty \quad \text{and} \quad 0 < \tilde{\varepsilon} \le \tilde{\sigma}(x) \le \tilde{M} < \infty, \qquad x \in \mathbb{R}.$$

Thus from Proposition 3.1 it satisfies the FKG inequality as in Theorem 7 of [3]. Here the same covariance inequality can be obtained without using the FKG inequality, and under weaker hypotheses.

Theorem 3.2. *Let $s, t \in [0,1]$ and assume that σ, $\tilde{\sigma}$ satisfy the condition*

$$\sigma_r(x)\,\tilde{\sigma}_r(y) \geq 0, \qquad x, y \in \mathbb{R}, \quad 0 \leq r \leq s \wedge t.$$

Then we have

$$\mathrm{Cov}(f(X_s), g(\tilde{X}_t)) \geq 0, \tag{12}$$

for all non-decreasing Lipschitz functions f, g.

Proof. From Proposition 1.2.3 and Theorem 2.2.1 of [9], we have $f(X_s) \in \mathrm{Dom}(D)$, $s \in [0,1]$, and

$$D_r f(X_s) = 1_{[0,s]}(r)\,\sigma_r(X_r)\,f'(X_s)\,e^{\int_r^s \alpha_u dW_u + \int_r^s (\beta_u - \frac{1}{2}\alpha_u^2)du}, \tag{13}$$

$r, s \in [0,1]$, where $(\alpha_u)_{u\in[0,1]}$ and $(\beta_u)_{u\in[0,1]}$ are uniformly bounded adapted processes. Hence we have

$$\mathbb{E}[D_r X_s \mid \mathcal{F}_r] = 1_{[0,s]}(r)\,\sigma_r(X_r)\,\mathbb{E}\Big[f'(X_s)\,e^{\int_r^s \alpha_u dW_u + \int_r^s (\beta_u - \frac{1}{2}\alpha_u^2)du}\ \Big|\ \mathcal{F}_r\Big],$$

$r, s \in [0,1]$. Similarly we show that $\mathbb{E}[D_r g(\tilde{X}_t) \mid \mathcal{F}_r]$ has the form

$$\mathbb{E}[D_r g(\tilde{X}_t) \mid \mathcal{F}_r] = 1_{[0,t]}(r)\,\tilde{\sigma}_r(\tilde{X}_r)\,\mathbb{E}\Big[g'(\tilde{X}_t)\,e^{\int_r^t \tilde{\alpha}_u dW_u + \int_r^t (\tilde{\beta}_u - \frac{1}{2}\tilde{\alpha}_u^2)du}\ \Big|\ \mathcal{F}_r\Big],$$

$r, t \in [0,1]$, and we conclude the proof from Lemma 2.1. □

Note that (12) has also been obtained for $s = t$ and $X = \tilde{X}$ in [7], Theorem 3.2, by semigroup methods. In this case it also follows by applying Corollary 1.4 of [5] in dimension one. The argument of [7] can in fact be extended to recover Theorem 3.2 as above. Also, (12) may also hold under local Lipschitz hypotheses on σ and $\tilde{\sigma}$, for example as a consequence of Corollary 4.2 of [1].

4. The discrete case

Let $\Omega = \{-1,1\}^{\mathbb{N}}$ and consider the family $(X_k)_{k\geq 1}$ of independent Bernoulli $\{-1,1\}$-valued random variables constructed as the canonical projections on Ω, under a measure $\mathbb{P}$ such that

$$p_n = \mathbb{P}(X_n = 1) \quad \text{and} \quad q_n = \mathbb{P}(X_n = -1), \qquad n \in \mathbb{N}.$$

Let $\mathcal{F}_{-1} = \{\varnothing, \Omega\}$ and

$$\mathcal{F}_n = \sigma(X_0, \ldots, X_n), \qquad n \in \mathbb{N}.$$

Consider the linear gradient operator D defined as

$$D_kF(\omega) = \sqrt{p_kq_k}\left(F((\omega_i 1_{\{i\neq k\}} + 1_{\{i=k\}})_{i\in\mathbb{N}}) - F(\omega_i 1_{\{i\neq k\}} - 1_{\{i=k\}})_{i\in\mathbb{N}}\right), \tag{14}$$

$k \in \mathbb{N}$. Recall the discrete Clark Formula, cf. Proposition 7 of [11]:

$$F = \mathbb{E}[F] + \sum_{k=0}^{\infty} \mathbb{E}[D_kF \mid \mathcal{F}_{k-1}]\, Y_k, \tag{15}$$

where

$$Y_k = 1_{\{X_k=1\}}\sqrt{\frac{q_k}{p_k}} - 1_{\{X_k=-1\}}\sqrt{\frac{p_k}{q_k}}, \quad k \in \mathbb{N},$$

defines a normalized i.i.d. sequence of centered random variables with unit variance. The Clark formula entails the following covariance identity, cf. Theorem 2 of [11]:

$$\mathrm{Cov}(F,G) = \mathbb{E}\left[\sum_{k=0}^{\infty} \mathbb{E}[D_kF \mid \mathcal{F}_{k-1}]\,\mathbb{E}[D_kG \mid \mathcal{F}_{k-1}]\right],$$

which yields a discrete time analog of Lemma 2.1.

Lemma 4.1. *Let $F, G \in L^2(\Omega)$ such that*

$$\mathbb{E}[D_kF \mid \mathcal{F}_{k-1}] \cdot \mathbb{E}[D_kG \mid \mathcal{F}_{k-1}] \geq 0, \quad k \in \mathbb{N}.$$

Then F and G are non-negatively correlated:

$$\mathrm{Cov}(F,G) \geq 0.$$

According to the next definition, a non-decreasing functional F satisfies $D_kF \geq 0$ for all $k \in \mathbb{N}$.

Definition 4.1. A random variable $F : \Omega \to \mathbb{R}$ is said to be non-decreasing if for all $\omega_1, \omega_2 \in \Omega$ we have

$$\omega_1(k) \leq \omega_2(k),\ \forall k \in \mathbb{N}, \quad \Rightarrow \quad F(\omega_1) \leq F(\omega_2).$$

The following result is then immediate from (14) and Lemma 4.1, and shows that the FKG inequality holds on Ω.

Proposition 4.1. *If $F, G \in L^2(\Omega)$ are non-decreasing then F and G are non-negatively correlated:*

$$\mathrm{Cov}(F,G) \geq 0.$$

Note however that the assumptions of Lemma 4.1 are actually weaker as they do not require F and G to be non-decreasing.

References

1. E. Alós, C. O. Ewald, "A note on the Malliavin differentiability of the Heston volatility", *Economics and Business Working Papers Series*, no. 880, Universitat Pompeu Fabra, 2005.
2. D. Bakry, D. Michel, "Sur les inégalités FKG" in *Séminaire de Probabilités, XXVI*, volume 1526 of *Lecture Notes in Math.*, Springer, Berlin, 1992, pp. 170–188.
3. D. Barbato, "FKG inequality for Brownian motion and stochastic differential equations", *Electron. Comm. Probab.* **10**, 7–16 (electronic) (2005).
4. C. M. Fortuin, P. W. Kasteleyn, J. Ginibre, "Correlation inequalities on some partially orderd sets", *Comm. Math. Phys.* **22**, 89–103 (1971).
5. I. Herbst, L. Pitt, "Diffusion equation techniques in stochastic monotonicity and positive correlations", *Probab. Theory Related Fields* **87**, 275–312 (1991).
6. C. Houdré, N. Privault, "Concentration and deviation inequalities in infinite dimensions via covariance representations", *Bernoulli* **8**(6), 697–720 (2002).
7. Y. Z. Hu, "Itô-Wiener chaos expansion with exact residual and correlation, variance inequalities", *J. Theoret. Probab.* **10**(4), 835–848 (1997).
8. D. Nualart, "Markov fields and transformations of the Wiener measure", in T. Lindstrøm, B. Øksendal, A. S. Üstünel, editors, *Proceedings of the Fourth Oslo-Silivri Workshop on Stochastic Analysis*, volume 8 of *Stochastics Monographs*, Oslo, 1993. Gordon and Breach.
9. D. Nualart, *The Malliavin Calculus and Related Topics.* Probability and its Applications, Springer-Verlag, 1995.
10. D. Ocone. "A guide to the stochastic calculus of variations", in H. Körezlioğlu, A. S. Üstünel, editors, *Stochastic Analysis and Related Topics*, volume 1316 of *Lecture Notes in Mathematics*, Silivri, 1988, Springer-Verlag.
11. N. Privault, W. Schoutens, "Discrete chaotic calculus and covariance identities", *Stochastics and Stochastics Reports*, **72**, 289–315 (2002).
12. A. S. Üstünel, *An introduction to analysis on Wiener space*, volume 1610 of *Lecture Notes in Mathematics*, Springer-Verlag, 1995.
13. S. Watanabe, *Lectures on Stochastic Differential Equations and Malliavin Calculus*, Tata Institute of Fundamental Research, 1984.
14. L. Wu, "A new modified logarithmic Sobolev inequality for Poisson point processes and several applications", *Probab. Theory Related Fields* **118**(3), 427–438 (2000).

YUTAO MA
Department of Mathematics, Wuhan University, 430072 Hubei, China
and Laboratoire de Mathématiques et Applications, Université de La Rochelle,
17042 La Rochelle, France
yutao.ma@univ-lr.fr

NICOLAS PRIVAULT
Laboratoire de Mathématiques et Applications, Université de La Rochelle,
17042 La Rochelle, France
nicolas.privault@univ-lr.fr

PATH-INTEGRAL ESTIMATES OF GROUND-STATE FUNCTIONALS

R. VILELA MENDES (*CMAF-UL, Lisboa*)

Based on a rigorous version of the path-integral ground-state method (currently used in molecular and condensed matter physics), a closed form representation of the ground-state functional is obtained. This provides a ground-state approximation scheme displaying non-perturbative features already at the lowest order.

1. Introduction

The path integral ground state (PIGS) method [1, 2] is currently used for practical calculations in molecules as well as in infinite systems like liquid and solid ^{4}He [3, 4]. It is based in an asymptotic estimate of the ground state wave function, namely

$$\psi_0(x) \propto \lim_{\tau\to\infty} \int dx'\, K(x, x', \tau)\, \psi_T(x'), \tag{1}$$

where x refers to the set of configuration space variables of the quantum many-body problem, $\psi_T(x')$ is a trial wave function and

$$K(x, x', \tau) = \left\langle x, e^{-\tau H} x' \right\rangle \tag{2}$$

is the imaginary time propagator for the Hamiltonian H, computed by dividing the time in small intervals $\Delta\tau$ and summing over polygonal paths ξ statistical sampled with a probability density

$$P(\xi) \propto \psi_T(x)\, \psi_T(x') \left\{ \prod_{j=0}^{N} K(x_j, x_{j+1}, \Delta\tau) \right\}. \tag{3}$$

Starting from a rigorous version of (1), the purpose of this paper is to obtain a ground state representation providing an approximation scheme with convergence properties better than a perturbative expansion. The approximation scheme uses a Gaussian approximation for paths centered at each point of the wave function, thus corresponding to a different expansion for each individual configuration point.

In Section 2, using conditions on the Hilbert spaces and Hamiltonians well established for finite-dimensional systems, several ground state representation formulas are established. They are rigorous versions of the formulae established in [5] by heuristic means and used in [6] to study nongeneric strata. In Section 3 a finite dimensional system is studied to exhibit the non-trivial nature of the leading order term in the ground state representation. Finally, in the conclusions, conditions are established which would allow for the extension of the results to infinite dimensional systems, in particular to the Yang-Mills theory in the Hamiltonian formulation.

2. The ground state path integral representation

Consider an Hilbert space $\mathcal{H}$ represented as $L^2(X, d\nu)$ for some measure space X with measure $d\nu$. In this space let H be an Hamiltonian

$$H = -\frac{1}{2}\Delta + V(x), \quad x \in X, \tag{4}$$

with $V(x)$ continuous and bounded from below and H essentially self-adjoint. Then the kernel $K_t(x, x')$ of e^{-tH} is strictly positive. This implies [7] the existence of a unique ground state Ω and for any $\psi \in L^2(X, d\nu)$ with $\|\psi\| \neq 0$ and $\psi \geq 0$

$$\left(\psi, e^{-tH}\Omega\right) = e^{-tE_0}(\psi, \Omega) > 0, \tag{5}$$

E_0 being the lowest eigenvalue of H.

From the spectral representation of H

$$e^{-tH} = \int e^{-tE} dP(E) \tag{6}$$

one obtains

$$\Omega = \frac{1}{(\Omega, \psi)} \lim_{t\to\infty} e^{tE_0} e^{-tH}\psi. \tag{7}$$

Under the stated conditions on H, one also has a Feynman-Kac representation for the kernel of e^{-tH},

$$e^{-tH}(x, x') = \int \exp\left(-\int_{-t}^{0} V(\gamma_s)\, ds\right) dW_{x,x'}^t, \tag{8}$$

$dW_{x,x'}^t$ being the Wiener measure for continuous paths on X going from x' to x in time t. Then

$$\Omega(x) = \frac{1}{(\Omega, \psi)} \lim_{t\to\infty} e^{tE_0} \int dx' \exp\left(-\int_{-t}^{0} V(\gamma_s)\, ds\right) dW_{x,x'}^t \psi(x'), \tag{9}$$

where γ_s is a Brownian path pinned down to x at time zero and to x' at time $-t$.

Equation (9) is a rigorous version of the path integral ground state method referred to in (1)–(3).

Practical calculations in quantum theory are made by some kind of approximation method, usually based on the calculation of the first few terms in a power series expansion. In the Feynman-Kac formula for the kernel (after the integrable part of the Hamiltonian is included in the measure), the perturbation expansion would be obtained by expanding the remaining exponential

$$\exp\left\{-\int V_r\left(\gamma_s\right)ds\right\}=\sum_{n=0}\frac{1}{n!}\left(-\int V_r\left(\gamma_s\right)ds\right)^n . \tag{10}$$

Usefulness of this expansion depends on the norm of V_r , that is on the strength of the coupling constant. On the other hand, the semiclassical approximation would be obtained by reintroducing the Planck constant h and expanding on it.

Here, instead, one uses an expansion of the argument of the exponential. For a strictly positive kernel, the ground state Ω may be chosen to be a strictly positive function. Therefore $\Omega^2(x)$ contains the same information as $\Omega(x)$. By a change $s \to -s$ in the time variable one converts the integral in (9) from $(-t, 0)$ to $(0, t)$. Then

$$\Omega^2(x) \propto \lim_{t\to\infty} e^{tE_0}\int dx'\delta\left(\gamma_0-x\right)\exp\left(-\int_{-t}^{t} V\left(\gamma_s\right)ds\right)dW^t_{x,x'}\psi\left(x'\right), \tag{11}$$

the Brownian paths being pinned to x at time zero.

Noticing that $\psi(x')$ is an arbitrary positive function, we may compute the integral (11) using a sequence of functions that are constant in successive larger domains (and zero outside). Choosing as normalization constant the same stochastic integral without the $\delta(\gamma_0 - x)$ one obtains

$$\Omega^2(x)=\frac{\int\delta\left(\gamma_0-x\right)\exp\left(-\int_{-\infty}^{\infty}V\left(\gamma_s\right)ds\right)dW^{\infty}}{\int\exp\left(-\int_{-\infty}^{\infty}V\left(\gamma_s\right)ds\right)dW^{\infty}}, \tag{12}$$

dW^∞ denoting the Wiener measure for Brownian paths for $t \in (-\infty, \infty)$.

To derive now from (12) a closed form expression, from which useful approximations may be obtained, one changes variables

$$\gamma_s = x + z_s \tag{13}$$

and adds a source term to the potential

$$V(x+z_s) \to V(x+z_s) - (z_s, J_s) \tag{14}$$

with $z_s \in S'$ and $J_s \in S$, S' and S being respectively the Schwartz distribution and test function spaces. Then, assuming analyticity of the potential and denoting by $G(x, z_s)$ the higher order terms,

$$\begin{aligned} &\exp\{-V(x+z_s) + (J_s, z_s)\} \\ &= \exp\left\{-V(x) - ((\Gamma(x) - J_s), z_s) - \tfrac{1}{2}(z_s, S(x) z_s) + G(x, z_s)\right\} \\ &= \exp\left\{G\left(x, \tfrac{\delta}{\delta J_s}\right)\right\} \exp\left\{-V(x) - ((\Gamma(x) - J_s), z_s) - \tfrac{1}{2}(z_s, S(x) z_s)\right\}, \end{aligned} \tag{15}$$

$\delta/\delta J_s$ denoting the Gateaux derivative.

$\delta(\gamma_0 - x) = \delta(z_0)$ may be represented by

$$\delta(z_0) = \frac{1}{2\pi} \int dy\, e^{-iy \int ds \delta(s) z_s}. \tag{16}$$

Substituting (15) and (16) in (12), one is left with Gaussian integrals over the Wiener measure. Computing these integrals one obtains

$$\Omega^2(x) = \left. \frac{e^{\int ds G\left(x, \frac{\delta}{\delta J_s}\right)} e^{\frac{1}{4}\int ds (\Gamma(x) - J_s) \frac{1}{-\frac{1}{2}\frac{\partial^2}{\partial s^2} + S(x)} (\Gamma(x) - J_s)} e^{\sqrt{2} L(x) \sqrt{S(x)} L(x)}}{\sqrt{\det \frac{1}{4\sqrt{\frac{1}{2} S(x)}}}\, e^{\int ds G\left(\frac{\delta}{\delta J_s}\right)} e^{\frac{1}{4}\int ds (\Gamma(x) - J_s) \frac{1}{-\frac{1}{2}\frac{\partial^2}{\partial s^2} + S(x)} (\Gamma(x) - J_s)}} \right|_{J_s = 0} \tag{17}$$

where

$$L(x) = -\frac{i\sqrt{2}}{8} \int ds \left\{ \frac{1}{\sqrt{S(x)}} e^{-|s|\sqrt{2S(x)}}, (\Gamma(x) - J_s) \right\}_+, \tag{18}$$

$\{\cdots\}_+$ meaning symmetrization. In the $J \to 0$ limit, $L(x)$ reduces to

$$-\frac{i}{4} \left\{ \frac{1}{S(x)}, \Gamma(x) \right\}_+. \tag{19}$$

Notice that the Gaussian integrations that were performed assume the positivity of the matrix $S(x)$.

By computing successive terms in $\exp\left(\int ds\, G(x, \delta/\delta J_s)\right)$ one obtains successive approximations for the ground state. Notice the highly non-trivial nature of the leading term

$$\Omega^2\,(x)_{(0)} =$$
$$\left(\det \sqrt{8S\,(x)}\right)^{\frac{1}{2}} \exp\left\{-\frac{1}{8\sqrt{2}}\left\{\Gamma\,(x)\,,\frac{1}{S\,(x)}\right\}_+ \sqrt{S\,(x)}\left\{\frac{1}{S\,(x)},\Gamma\,(x)\right\}_+\right\}. \tag{20}$$

3. A finite-dimensional example

Consider the Hamiltonian

$$H = -\frac{1}{2}\frac{d}{dy^2} + V\,(y) \tag{21}$$

with

$$V\,(y) = \frac{1}{2}y^2 - \mu y \tanh \mu y. \tag{22}$$

In a neighborhood of the origin this potential is qualitatively similar to the quartic anharmonic potential. However to illustrate the approximation scheme developed in the previous section it is a more convenient potential because its ground state is exactly known

$$\Omega\,(x) = \exp\left(\frac{-\,(x-\mu)^2}{2}\right) + \exp\left(\frac{-\,(x+\mu)^2}{2}\right). \tag{23}$$

Making in (22) the change of variables

$$y = x + z \tag{24}$$

and computing the linear and quadratic terms in an expansion of the potential around x, one obtains

$$\begin{aligned}\Gamma\,(x) &= \mu - \mu^2 x\left(1-\tanh^2(\mu x)\right) - \mu\tanh(\mu x)\\ S\,(x) &= \frac{1}{2} - \mu^3 x\left(\tanh^3(\mu x) - \tanh(\mu x)\right) - \mu^2\left(1-\tanh^2(\mu x)\right)\end{aligned}$$

with a leading term approximation for the ground state

$$\Omega_{(0)} = (8S\,(x))^{1/8}\exp\left(-\frac{1}{4\sqrt{2}}\frac{\Gamma^2\,(x)}{S^{3/2}\,(x)}\right). \tag{25}$$

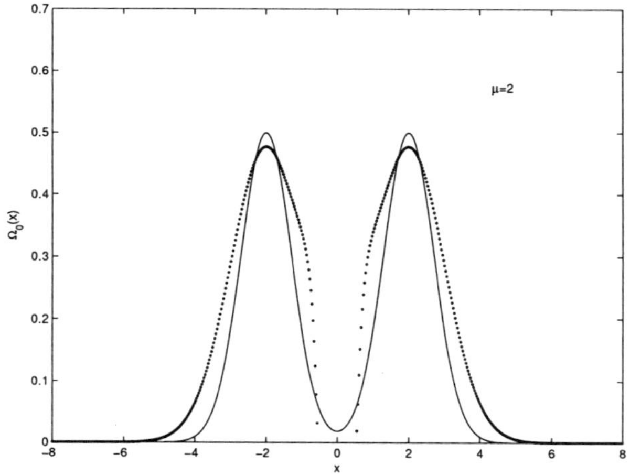

Figure 1. Exact ground state and leading order approximation

In the figure (for a coupling constant value $\mu = 2$) we compare the exact ground (23) (continuous line) with the approximation (25) (dotted plot). The region suppressed, near $x = 0$, corresponds to the region where $S(x) < 0$, outside the validity of the Gaussian integrations leading to (25). Notice the good qualitative agreement of $\Omega_{(0)}$ with the exact solution, whereas from an expansion in the coupling constant

$$H = -\frac{1}{2}\frac{d}{dy^2} + \left(\frac{1}{2} - \mu^2\right)y^2 + \frac{\mu^4}{3}y^4 - \frac{2}{15}\mu^6 y^6 + \cdots \qquad (26)$$

one concludes that, for this value of the coupling constant, a perturbative treatment would be hopeless.

4. Conclusions

1 – The reason why one has obtained results beyond the perturbation expansion is because, for the ground state wave function, the expansion is performed around each x. Therefore one is effectively dealing with a different expansion at each point. The improved convergence properties of expansions when the expansion center changes at each step of the calculations is a familiar feature for example in the Newton method or in KAM calculations.

2 – When X is a finite dimensional space ($X \subset \mathbb{R}^d$) and under the stated conditions on the Hilbert space $\mathcal{H}$, namely representatibility as $L^2(X, d\nu)$, a real potential bounded from below and an essentially self-adjoint Hamiltonian, the ground state representation formulae (11) or (12) are valid. An important question for applications to Yang-Mills, Chern-Simons or string theories is whether the results can be extended to those infinite dimensional cases. Of particular relevance would be to extend the rigorous formulation to the Yang-Mills case, in particular to establish a rigorous foundation for the mass gap results obtained in [5].

A classical Yang-Mills theory consists of four basic objects:

(i) A principal fiber bundle $P(M, G)$ with structural group G and projection $\pi : P \to M$, the base space M being an oriented Riemannian manifold.
(ii) An affine space $\mathcal{C}$ of connections ω on P, modelled by a vector space $\mathcal{A}$ of 1-forms on M with values on the Lie algebra $\mathcal{G}$ of G.
(iii) The space of differentiable sections of P, called the *gauge group* $\mathcal{W}$.
(iv) A $\mathcal{W}$−invariant functional (the Lagrangian) $\mathcal{L} : \mathcal{A} \to \mathbb{R}$.

Choosing a reference connection, the affine space of connections on P may be modelled by a vector space of $\mathcal{G}$-valued 1-forms $(C^\infty(\Lambda^1 \otimes \mathcal{G}))$. Likewise the curvature F is identified with an element of $(C^\infty(\Lambda^2 \otimes \mathcal{G}))$. The *configuration space of the gauge theory* is the quotient space $\mathcal{A}/\mathcal{W}$.

For an Hamiltonian formulation of Yang-Mills, with $A^\mu = A^\mu_a t_a$ ($\{t_a\}$ a basis for the Lie algebra), one takes

$$\begin{aligned} &A^i_a \\ &E^i_a = F^{i0}_a = \partial^i A^0_a - \partial^0 A^i_a - f_{bca} A^i_b A^0_c \end{aligned} \tag{27}$$

as canonical variables. To extend the ground state discussion to the Yang-Mills case one writes the Yang-Mills first-order action as

$$I = \frac{2}{g^2} \int d^4x \,\mathrm{Tr} \left\{ \partial^0 A \cdot E + \frac{1}{2}\left(E^2 + B^2\right) - A^0 \left(\nabla \cdot E + [A, E]\right) \right\} \tag{28}$$

with $B_a = -\frac{1}{2}\epsilon_{ijk} F^{jk}_a$. Then the Hamiltonian is

$$H = \int d^3x \sum_a \left(E^2_a + B^2_a\right) \tag{29}$$

and A^0 being a Lagrange multiplier, the constraint is

$$\Gamma(x) = \nabla \cdot E + [A, E] = D \cdot E = 0 \tag{30}$$

and the canonical brackets

$$\left\{A_a^i(x), E_b^j(y)\right\}_{x^0=y^0} = \delta^{ij}\,\delta_{ab}\,\delta^3(x-y)\,. \tag{31}$$

The constraint (Gauss's law) simply means that the allowed physical states should be gauge invariant.

For path integral calculations a measure in $\mathcal{A}/\mathcal{W}$ would be desirable, but no such measure has been found for Sobolev connections. Therefore it is more convenient to work in a space of *generalized connections* $\overline{\mathcal{A}}$, defining parallel transports on piecewise smooth paths as simple homomorphisms from the paths on M to the group G, without a smoothness assumption[8]. The same applies to the generalized gauge group $\overline{\mathcal{W}}$. Then, there is in $\overline{\mathcal{A}}/\overline{\mathcal{W}}$ an induced Haar measure, the Ashtekar-Lewandowski measure[9]–[10]. Sobolev connections are a dense zero measure subset of the generalized connections [11].

The extension of ground state representation, derived in the Section 2, to this case (in the continuum), depends on:

(i) the possibility of representing the Hilbert space $\mathcal{H}$ of states as $L^2\left(\overline{\mathcal{A}}, d\mu_A\right)$ for some measure μ_A,

(ii) the existence of a Wiener measure on paths on $\overline{\mathcal{A}}$,

(iii) a Feynman-Kac formula in this infinite-dimensional setting.

A natural setting for these purposes will be the construction of a Gelfand triplet

$$(S_{\mathcal{A}}) \subset L^2(\mathcal{A}, B, \mu_A) \subset (S_{\mathcal{A}})^*\,,$$

μ_A being the Ashtekar-Lewandowski measure.

References

1. D. M. Ceperley, *Rev. Mod. Phys.* **67**, 279 (1995).
2. A. Sarsa, K. E. Schmidt, W. R. Magro, "A path integral ground state method", *J. Chem. Phys.* **113**, 1366 (2000).
3. J. E. Cuervo, P.-N. Roy, M. Boninsegni, "Path integral ground state with a fourth-order propagator: Application to condensed helium", *J. Chem. Phys.* **122**, 114504 (2005).
4. W. Purwanto, S. Zhang, "Quantum Monte Carlo for the ground state of many-boson systems", *Phys. Rev. E* **70**, 056702 (2004).
5. R. Vilela Mendes, "Stochastic processes and the non-perturbative structure of the QCD vacuum", *Z. Phys. C — Particles and Fields* **54**, 273–281 (1992).
6. R. Vilela Mendes, "Stratification of the orbit space in gauge theories. The role of nongeneric strata", *J. Phys. A: Math. Gen.* **37**, 11485–11498 (2004).
7. L. Gross, "Existence and uniqueness of physical ground states", *J. Funct. Anal.* **10**, 52–109 (1972).

8. A. Ashtekar, C. J. Isham, "Representations of the holonomy algebras of gravity and nonabelian gauge theories", *Class. Quant. Grav.* **9**, 1433–1468 (1992).
9. A. Ashtekar, J. Lewandowski, "Differential geometry on the space of connections via graphs and projective limits", *J. Geom. Phys.* **17**, 191–230 (1995).
10. A. Ashtekar, J. Lewandowski, "Projective techniques and functional integration for gauge theories", *J.Math. Phys.* **36**, 2170–2191 (1995).
11. D. Marolf, J. M. Mourão, "On the support of the Ashtekar-Lewandowski measure", *Commun. Math. Phys.* **170**, 583–606 (1995).

R. Vilela Mendes
CMAF, Complexo Interdisciplinar, Universidade de Lisboa,
Av. Gama Pinto, 2, 1649-003 Lisboa, Portugal
vilela@cii.fc.ul.pt

A REPRESENTATION THEOREM AND A SENSITIVITY RESULT FOR FUNCTIONALS OF JUMP DIFFUSIONS

GIULIA DI NUNNO (*University of Oslo*),
BERNT ØKSENDAL (*University of Oslo / NHH, Bergen*)

We use white noise calculus for Lévy processes to obtain a representation formula for the functionals of a jump diffusion. Then we use this to find an explicit formula for the Donsker delta function of a jump diffusion and we suggest its application to sensitivity analysis in mathematical finance for the computation of the Greeks.

1. Introduction

A difficult, but crucial, task in the analysis of option prices is the prediction of their variation. To this aim it is important to locate which are the factors contributing to the fluctuation of prices and their effect. The *sensitivity analysis* is carried over the parameters appearing in the models for the price dynamics and the so-called *Greeks* represent a form of measure for the price sensitivity to some factors. For example, the "delta" is related to the initial price of the option, the "theta" is related to the time until maturity, the "rho" to the interest rate, the "vega" is the sensitivity to the volatility, etc. Efficient techniques for the computation of the Greeks rely on numerical finite difference methods and simulation. See [16], [14], for example and references therein.

However, too often some restriction on the regularity of the price processes has to be imposed. In the recent years high attention was dedicated to finding more efficient and more general methods to apply numerics and simulation for the computation of the Greeks. The papers [13] and [12] proved that, with a preliminary application of sophisticated tools of stochastic analysis, some better formulae could be derived which would ease a direct application of Monte Carlo simulation. Their method, based on Malliavin calculus, applies to price dynamics driven by Brownian motion only. See also [5], [15], [24], [26], for example, and references therein.

Several forms of generalization or extension to include dynamics driven by Poisson processes or combinations of independent Brownian motions and Poisson processes have been suggested. We can refer to [4], [8], [11], [32], for example.

In this paper we present a representation formula for functionals of jump diffusions (see Theorem 3.1) which, if applied to the sensitivity analysis context, gives a computational efficient formula for the Greek "delta". We frame our method in the setting of white noise analysis. A short introduction to this framework with the preliminary results is given in Section 2.

Section 3 presents the representation formula for functionals of a jump diffusion. Moreover, we apply this result to give an explicit representation of the Donsker delta function. Our approach is in the same line as [27]. This results gain importance in view of the applications of the Donsker delta function for the computation of hedging portfolios in mathematical finance. See [2] for the Brownian motion setting and [9] for the pure jump Lévy processes case.

Section 4 is dedicated to the sensitivity analysis.

2. Framework

Let $(\Omega, \mathcal{F}, P)$ be a complete probability space and $L_2(P)$ the standard (complex) L_2-space of the random variables ξ with finite norm $\|\xi\| := (E|\xi|^2)^{1/2} < \infty$. On the given probability space we consider the real *Lévy process* $\eta(t)$, $t \geq 0$, characterized by the Kolmogorov-deFinetti law

$$\log E\big[e^{iu\eta(t)}\big] = t\bigg[i\alpha u - \tfrac{1}{2}\sigma^2 u^2 + \int_{\mathbb{R}_0} (e^{iuz} - 1 - iuz)\,\nu(dz)\bigg], \quad u \in \mathbb{R},$$

where $\alpha \in \mathbb{R}$, $\sigma^2 > 0$ are constants and $\nu(dz)$, $z \in \mathbb{R}_0$, is a σ-finite Borel measure on $\mathbb{R}_0 := \mathbb{R} \setminus \{0\}$. Note that

$$\int_{\mathbb{R}_0} z^2 \nu(dz) < \infty\,,$$

namely we assume that $\eta(t) \in L_2(P)$ for all $t \geq 0$.

Throughout this paper we will always consider the càdlàg modification $\eta(t)$, $t \geq 0$, of the stochastic process above. We can refer to [6], [7], [33], for example, for general and detailed information about Lévy processes. In particular we recall that, for every t, the random variable $\eta(t)$ admits a representation in the form

$$\eta(t) = \alpha t + \sigma B(t) + \int_0^t \int_{\mathbb{R}_0} z\tilde{N}(ds, dz), \tag{1}$$

where the standard Brownian motion $B(t)$, $t \geq 0$, and the compensated Poisson random measure

$$\tilde{N}(dt, dz) := N(dt, dz) - \nu(dz)dt, \quad t \geq 0,\ z \in \mathbb{R},$$

are independent — cf. [20].

Inspired by the stochastic integral representation (1) it is natural to consider stochastic processes $\xi(t)$, $t \geq 0$, of the form

$$\xi(t) = \xi(0) + \int_0^t \alpha(s)\, ds + \int_0^t \beta(s)\, dB(s) + \int_0^t \int_{\mathbb{R}_0} \gamma(s,z)\, \tilde{N}(ds,dz), \quad (2)$$

where $\alpha(t)$, $\beta(t)$ and $\gamma(t,z)$, $t \geq 0$, $z \in \mathbb{R}_0$, are *deterministic* functions satisfying

$$\int_0^\infty \left[|\alpha(t)| + \beta^2(t) + \int_{\mathbb{R}_0} \gamma^2(t,z)\, \nu(dz) \right] dt < \infty . \quad (3)$$

On the other side, in line with the approach suggested in [3], we could consider a representation of type (2) embedded in a multidimensional framework as follows. Let us consider the probability space $(\Omega, \mathcal{F}, P)$ as a product of two complete probability spaces, i.e.,

$$\Omega = \Omega_1 \times \Omega_2, \quad \mathcal{F} = \mathcal{F}_1 \otimes \mathcal{F}_2 \quad P = P_1 \otimes P_2. \quad (4)$$

In such a framework we could consider stochastic processes $\xi(t)$, $t \geq 0$, on $(\Omega, \mathcal{F}, P)$ such that

$$\begin{aligned} \xi(t, \omega_1, \omega_2) = y &+ \int_0^t \alpha(s)\, ds + \int_0^t \beta(s)\, dB(s, \omega_1) \\ &+ \int_0^t \int_{\mathbb{R}_0} \gamma(s,z)\, \tilde{N}(ds, dz, \omega_2) \end{aligned} \quad (5)$$

for $y \in \mathbb{R}$ constant and $\alpha(t), \beta(t)$ and $\gamma(t,z)$, $t \geq 0$, $z \in \mathbb{R}_0$, deterministic functions satisfying (3).

We equip the probability space $(\Omega_1, \mathcal{F}_1, P_1)$ with the filtration $\mathcal{F}_t^1$, $t \geq 0$, $(\mathcal{F}_\infty^1 = \mathcal{F}_1)$ generated by $B(t)$, $t \geq 0$, augmented of all P_1-null sets and the space $(\Omega_2, \mathcal{F}_2, P_2)$ with the filtration $\mathcal{F}_t^2$, $t \geq 0$ $(\mathcal{F}_\infty^2 = \mathcal{F}_2)$ generated by the values of $\tilde{N}(dt,dz)$, $t \geq 0$, $z \in \mathbb{R}_0$, augmented of all P_2-null sets. Then on the product $(\Omega, \mathcal{F}, P)$ we fix the filtration

$$\mathcal{F}_t := \mathcal{F}_t^1 \otimes \mathcal{F}_t^2, \quad t \geq 0.$$

In the sequel we apply white noise analysis and techniques. Thus we choose to set $(\Omega_1, \mathcal{F}_1, P_1)$ to be a *Gaussian white noise probability space* and $(\Omega_2, \mathcal{F}_2, P_2)$ a *Poissonian white noise probability space.*

General references to white noise theory for Gaussian processes are e.g. [17], [18], [19], [25], [28]. As for a white noise theory to non-Gaussian analysis we can refer to e.g. [1], [10], [22], [23], [29], [31]. In order to keep this presentation moderate in size we recall here only the Poisson white noise framework in the approach and notation of [10] and [29].

To ease the notation we drop the index of $(\Omega_2, \mathcal{F}_2, P_2)$ and we write $(\Omega_2, \mathcal{F}_2, P_2) = (\Omega, \mathcal{F}, P)$ from now up to the end of this section.

From now on we assume that for every $\varepsilon > 0$ there exists $\rho > 0$ such that

$$\int_{\mathbb{R}\setminus(-\varepsilon,\varepsilon)} e^{\rho|z|}\nu(dz) < \infty. \tag{6}$$

This condition implies that the polynomials are dense in $L_2(\mu)$ where $\mu(dz) = z^2\nu(dz)$. It also guarantees that the measure ν integrates all polynomials of degree greater than or equal to 2.

Let $\mathcal{A}$ denote the set of all multi-indices $\alpha = (\alpha_0, \alpha_1, \ldots)$ which have only finitely many non-zero values $\alpha_i \in \mathbb{N} \setminus \{0\}$. In the space $L_2(\Omega, \mathcal{F}, P) = L_2(\Omega_2, \mathcal{F}_2, P_2)$ we construct the orthogonal basis K_α, $\alpha \in \mathcal{A}$, as follows.

First of all we consider the orthonormal basis φ_i, $i \in \mathbb{N}$, in $L_2(\lambda)$ constituted by the Laguerre functions (order 1/2). Here and in the sequel $\lambda(dt) = dt$ denotes the Lebesgue measure on the real line. Moreover we take an orthonormal basis ψ_j, $j \in \mathbb{N}$, in $L_2(\nu)$ of polynomial type. See e.g. [29] for further details.

Then we can consider the products

$$\zeta_k(t, z) = \varphi_i(t)\,\psi_j(z) \tag{7}$$

for $k = k(i,j)$ as a bijective mapping $k : \mathbb{N} \times \mathbb{N} \longrightarrow \mathbb{N}$ (e.g. the diagonal counting of the Cartesian product $\mathbb{N} \times \mathbb{N}$).

For any $\alpha \in \mathcal{A}$ with $\max\{i : \alpha_i \neq 0\} = j$ and $|\alpha| := \sum_i \alpha_i = m$, we can define

$$\zeta^{\otimes\alpha}\big((t_1, z_1), \ldots, (t_m, z_m)\big) := \zeta_1^{\otimes\alpha_1} \otimes \cdots \otimes \zeta_j^{\otimes\alpha_j}\big((t_1, z_1), \ldots, (t_m, z_m)\big) =$$
$$\zeta_1(t_1, z_1) \cdots \zeta_1(t_{\alpha_1}, z_{\alpha_1}) \cdots \zeta_j(t_{\alpha_1+\cdots+\alpha_{j-1}+1}, z_{\alpha_1+\cdots+\alpha_{j-1}+1}) \cdots \zeta_j(t_m, z_m)$$

and $\zeta^{\otimes 0} = 1$. Moreover, we denote the corresponding *symmetrized tensor product* by $\zeta^{\hat{\otimes}\alpha}$. We can now construct an *orthogonal basis* K_α, $\alpha \in \mathcal{A}$, in $L_2(P)$ as follows:

$$K_\alpha := I_{|\alpha|}(\zeta^{\hat{\otimes}\alpha}), \quad \alpha \in \mathcal{A}, \tag{8}$$

where

$$I_n(f) = n! \int_0^\infty \int_{\mathbb{R}} \cdots \int_0^{t_2} \int_{\mathbb{R}} f(t_1, z_1, \ldots, t_n, z_n)\, \widetilde{N}(dt_1, dz_1) \cdots \widetilde{N}(dt_n, dz_n)$$

is the Itô iterated integral with respect to the centered Poisson stochastic measure. See [20]. Here $f \in L_2(\lambda \times \nu)^n$ is symmetric in the n pairs $(t_1, z_1), \dots, (t_n, z_n)$. Note that

$$E\big[I_m(g) \cdot I_n(f)\big] = 0, \quad m \neq n,$$

and

$$E\big[I_n(f)^2\big] = n! \, \|f\|^2_{L_2(\lambda\times\nu)^n}$$

for all symmetric $g \in L_2(\lambda \times \nu)^m$ and $f \in L_2(\lambda \times \nu)^n$ $(m, n \in \mathbb{N})$.

Hence every $\xi \in L_2(P)$ admits the chaos expansion

$$\xi = \sum_{\alpha\in\mathcal{A}} c_\alpha K_\alpha \quad (c_\alpha \in \mathbb{R}) \tag{9}$$

and

$$\|\xi\|^2_{L_2(P)} = \sum_{\alpha\in\mathcal{A}} c_\alpha^2 \, \|K_\alpha\|^2_{L_2(P)} = \sum_{\alpha\in\mathcal{A}} c_\alpha^2 \, \alpha! \tag{10}$$

where $\alpha! := \alpha_1! \, \alpha_2! \dots$ for $\alpha = (\alpha_1, \alpha_2, \dots) \in \mathcal{A}$.

Thanks to these chaos expansions we can characterize the following spaces and chain of embeddings.

By $(\mathcal{S})_\rho$ $(0 \leq \rho \leq 1)$ we denote the space of all random variables $f = \sum_{\alpha\in\mathcal{A}} c_\alpha K_\alpha \in L_2(P)$ such that

$$\|f\|^2_{\rho,k} = \sum_{\alpha\in\mathcal{A}} (\alpha!)^{1+\rho} \, c_\alpha^2 (2\mathbb{N})^{k\alpha} < \infty \qquad \text{for all } k \in \mathbb{N},$$

where $(2\mathbb{N})^{k\alpha} := (2\cdot 1)^{k\alpha_1}(2\cdot 2)^{k\alpha_2} \cdots (2\cdot j)^{k\alpha_j}$ if $j = \max\{i : \alpha_j \neq 0\}$. And by $(\mathcal{S})_{-\rho}$ we denote the space of all random variables $F = \sum_{\alpha\in\mathcal{A}} c_\alpha K_\alpha \in L_2(P)$ such that

$$\|F\|^2_{-\rho,-k} = \sum_{\alpha\in\mathcal{A}} (\alpha!)^{1-\rho} \, c_\alpha^2 (2\mathbb{N})^{-k\alpha} < \infty \qquad \text{for some } k \in \mathbb{N}.$$

The subspaces $(\mathcal{S})_\rho$ and $(\mathcal{S})_{-\rho}$ are respectively equipped with the projective topology and the inductive topology induced by the above seminorms. Note that for any $F = \sum_{\alpha\in\mathcal{A}} a_\alpha K_\alpha \in (\mathcal{S})_{-\rho}$ and $f = \sum_{\alpha\in\mathcal{A}} b_\alpha K_\alpha \in (\mathcal{S})_\rho$ the action

$$\langle F, f\rangle := \sum_{\alpha\in\mathcal{A}} a_\alpha \, b_\alpha \, \alpha!$$

is well-defined and thus the space $(\mathcal{S})_{-\rho}$ is the dual of $(\mathcal{S})_\rho$, i.e., $(\mathcal{S})_{-\rho} = (\mathcal{S})^*_\rho$. We remark that, for $\rho = 0$, the spaces $(\mathcal{S}) := (\mathcal{S})_0$ and $(\mathcal{S})^* = (\mathcal{S})^*_0 =$

$(\mathcal{S})_{-0}$ appear respectively as a Lévy version for the *Hida test function space* and *Hida distribution space* for pure jump Lévy processes. See e.g. [18], [19], [25], [28]. For $\rho = 1$, the spaces $(\mathcal{S})_1$ and $(\mathcal{S})_{-1}$ are the Lévy version of the *Kondratiev test function space* and the *Kondratiev distribution space* respectively. See [21] and also [19], for example.

The following relationships hold true

$$(\mathcal{S})_1 \subset (\mathcal{S})_\rho \subset (\mathcal{S}) \subset L_2(P) \subset (\mathcal{S})^* \subset (\mathcal{S})_{-\rho} \subset (\mathcal{S})_{-1}.$$

The relevance of these spaces will be clarified in the sequel. For instance $(\mathcal{S})^*$ is rich enough to contain the white noise of the centered Poisson stochastic measure and of the pure jump Lévy process as its elements. In fact, let us consider the random variable $\xi = \widetilde{N}(t, B) \in L_2(P)$, for any Borel set $B \in \mathcal{B}(\mathbb{R}\backslash\{0\})$, then the following chaos expansion can be written:

$$\widetilde{N}(t, B) = \sum_{i,j\geq 1} \int_0^t \int_B \varphi_i(s)\, \psi_j(z)\, \nu(dz)\, ds \cdot K_{\epsilon^{k(i,j)}}$$

via the application of the orthogonal basis in $L_2(\lambda)$ and $L_2(\nu)$ and the diagonal counting — cf. (7). Moreover, for any $k \in \mathbb{N}$, the multi-index $\epsilon^k = (\epsilon_1^k, \epsilon_2^k, \ldots)$ is defined by

$$\epsilon_i^k := \begin{cases} 1, & i = k \\ 0, & \text{otherwise.} \end{cases}$$

Here we refer to [29] for all the details. Then we can define the *white noise for the centered Poisson stochastic measure* as the element

$$\overset{\bullet}{\widetilde{N}}(t, z) := \sum_{i,j\geq 1} \varphi_i(t)\, \psi_j(z) \cdot K_{\epsilon^{k(i,j)}} \tag{11}$$

in $(\mathcal{S})^*$, for almost all $t \geq 0$, $z \in \mathbb{R}$. Naturally it appears as the Radon-Nikodym derivative

$$\overset{\bullet}{\widetilde{N}}(t, z) = \frac{\widetilde{N}(dt, dz)}{dt \times \nu(dz)} \quad \text{in } (\mathcal{S})^*.$$

The *Lévy-Wick product* $F \diamond G$ of two elements $F = \sum_{\alpha\in\mathcal{A}} a_\alpha K_\alpha$ and $G = \sum_{\beta\in\mathcal{A}} b_\beta K_\beta$ in $(\mathcal{S})_{-1}$ is defined by

$$F \diamond G = \sum_{\alpha,\beta\in\mathcal{A}} a_\alpha\, b_\beta\, K_{\alpha+\beta}. \tag{12}$$

It can be shown that the spaces $(\mathcal{S})_1, (\mathcal{S}), (\mathcal{S})^*$ and $(\mathcal{S})_{-1}$ are closed under Wick products.

One of the useful features of the Wick product is the following relationship within Itô stochastic integration and Bochner integration:

$$\int_0^t \int_{\mathbb{R}} Y(s,z)\,\widetilde{N}(ds,dz) = \int_0^t \int_{\mathbb{R}} Y(s,z) \diamond \overset{\bullet}{\widetilde{N}}(s,z)\,\nu(dz)\,ds. \tag{13}$$

We also mention that for all $F \in (\mathcal{S})_{-1}$ one can define the *Wick exponential* $\exp^\diamond F \in (\mathcal{S})_{-1}$ by

$$\exp^\diamond F := \sum_{n=0}^{\infty} \frac{1}{n!} F^{\diamond n} \tag{14}$$

and the following property

$$E(\exp^\diamond F) = \exp\{EF\} \tag{15}$$

holds true.

3. A representation theorem for functionals of a class of jump diffusions

Let $\xi(t) = \xi^y(t)$, $t \in [0,T]$, be a stochastic process on $(\Omega, \mathcal{F}, P)$ — cf. (4), of the form

$$\begin{cases} d\xi(t) = \alpha(t)\,dt + \beta(t)\,dB(t) + \int_{\mathbb{R}} \gamma(t,z)\,\tilde{N}(dt,dz), & t \in [0,T], \\ \xi(0) = y \in \mathbb{R} \end{cases} \tag{16}$$

where $\alpha(t), \beta(t)$ and $\gamma(t,z)$, $t \in [0,T]$, $z \in \mathbb{R}$, are deterministic functions satisfying (3) — cf. (5). For $\lambda \in \mathbb{R}$ define

$$Y(t) = \exp(\lambda \xi(t)), \quad t \in [0,T]. \tag{17}$$

Then by the Itô formula (see e.g. [30, Chapter 1]) we have

$$\begin{aligned} dY(t) = Y(t^-)\Big[&\big(\lambda\alpha(t) + \tfrac{1}{2}\lambda^2\beta^2(t)\big)\,dt + \lambda\beta(t)\,dB(t) \\ &+ \int_{\mathbb{R}} \{\exp(\lambda\gamma(t,z)) - 1 - \lambda\gamma(t,z)\}\,\nu(dz)\,dt \\ &+ \int_{\mathbb{R}} \{\exp(\lambda\gamma(t,z)) - 1\}\tilde{N}(dt,dz)\Big]. \end{aligned} \tag{18}$$

Using white noise notation and Wick calculus this can be written

$$\begin{aligned}\frac{dY(t)}{dt} = Y(t^-) \diamond \Big\{ & \lambda\alpha(t) + \tfrac{1}{2}\lambda^2\beta^2(t) + \lambda\beta(t)\dot{B}(t) \\ & + \int_{\mathbb{R}} \big[\exp\{\lambda\gamma(t,z)\} - 1 - \lambda\gamma(t,z)\big]\,\nu(dz) \\ & + \int_{\mathbb{R}} \{\exp\{\lambda\gamma(t,z)\} - 1\}\,\overset{\bullet}{\tilde{N}}(t,z)\,\nu(dz)\Big\}; \quad Y(0) = e^{\lambda y}.\end{aligned} \tag{19}$$

The solution of (19) is, using Wick calculus in $(\mathcal{S})^*$,

$$\begin{aligned}Y(t) = Y(0)\exp^{\diamond}\Big\{ & \int_0^t \Big\{\lambda\alpha(s) + \tfrac{1}{2}\lambda^2\beta^2(s) \\ & + \int_{\mathbb{R}} \{e^{\lambda\gamma(s,z)} - 1 - \lambda\gamma(s,z)\}\,\nu(dz)\Big\}\,ds \\ & + \int_0^t \lambda\beta(s)\,dB(s) + \int_0^t\int_{\mathbb{R}} \{e^{\lambda\gamma(s,z)} - 1\}\tilde{N}(ds,dz)\Big\}.\end{aligned} \tag{20}$$

Comparing (20) with (17) we get the following formula for the Wick exponential.

Lemma 3.1. *With $\xi(t)$ as in (16) and $\lambda \in \mathbb{R}$ we have*

$$\begin{aligned}e^{\lambda\xi(t)} = e^{\lambda y}\exp^{\diamond}\Big\{ & \int_0^t \Big[\lambda\alpha(s) + \tfrac{1}{2}\lambda^2\beta^2(s) \\ & + \int_{\mathbb{R}} \big[e^{\lambda\gamma(s,z)} - 1 - \lambda\gamma(s,z)\big]\,\nu(dz)\Big]\,ds \\ & + \int_0^t \lambda\beta(s)\,dB(s) + \int_0^t\int_{\mathbb{R}} \big[e^{\lambda\gamma(s,z)} - 1\big]\,\tilde{N}(ds,dz)\Big\}.\end{aligned} \tag{21}$$

Using this we obtain the following result:

Theorem 3.1 (Representation theorem for functionals of a jump diffusion). *Let $g : \mathbb{R} \to \mathbb{R}$ be a function with Fourier transform*

$$\hat{g}(\lambda) = \frac{1}{2\pi}\int_{\mathbb{R}} e^{-i\lambda x} g(x)\,dx, \quad \lambda \in \mathbb{R},$$

and satisfying the Fourier inversion property

$$g(u) = \int_{\mathbb{R}} e^{i\lambda u} \hat{g}(\lambda)\, d\lambda, \quad u \in \mathbb{R}.$$

Then

$$g(\xi^y(t)) = \int_{\mathbb{R}} \hat{g}(\lambda) \exp^\diamond \{X_\lambda^y(t)\}\, d\lambda, \quad t \in [0,T], \tag{22}$$

where

$$\begin{aligned} X_\lambda^y(t) = i\lambda y &+ \int_0^t i\lambda\beta(s)\, dB(s) + \int_0^t \int_{\mathbb{R}} \left[e^{i\lambda\gamma(s,z)} - 1\right] \tilde{N}(ds,dz) \\ &+ \int_0^t \Big\{ i\lambda\alpha(s) - \tfrac{1}{2}\lambda^2\beta^2(s) \\ &\qquad + \int_{\mathbb{R}} \left[e^{i\lambda\gamma(s,z)} - 1 - i\lambda\gamma(s,z)\right] \nu(dz) \Big\}\, ds, \quad t \in [0,T]. \end{aligned} \tag{23}$$

Proof. Applying (21) with $i\lambda$ instead of λ we get

$$g(\xi^y(t)) = \frac{1}{2\pi} \int_{\mathbb{R}} e^{i\lambda\xi(t)} \hat{g}(\lambda)\, d\lambda = \frac{1}{2\pi} \int_{\mathbb{R}} \hat{g}(\lambda) \exp^\diamond (X_\lambda^y(t))\, d\lambda.$$

□

Corollary 3.1. *Let g be a real function as in Theorem 3.1. Then we have*

$$E[g(\xi^y(t))] = \int_{\mathbb{R}} \hat{g}(\lambda) \cdot \exp(i\lambda y + G_\lambda(t))\, d\lambda, \tag{24}$$

where

$$G_\lambda(t) = \int_0^t \Big\{ i\lambda\alpha(s) - \tfrac{1}{2}\lambda^2\beta^2(s) + \int_{\mathbb{R}} \left(e^{i\lambda\gamma(s,z)} - 1 - i\lambda\gamma(s,z)\right) \nu(dz) \Big\}\, ds. \tag{25}$$

Proof. This follows from Theorem 3.1 plus the fact that (see (15))

$$E[\exp^\diamond X_\lambda^y(t)] = \exp(E[X_\lambda^y(t)]).$$

□

In the last part of this section we obtain an explicit formula for the Donsker delta function of $\xi(t) = \xi^y(t)$, $t \geq 0$. This is derived as an application of Theorem 3.1.

The Donsker delta function is a generalized white noise functional, we can refer e.g. [17], [18], [25], for general information. Here we give its definition within the white noise framework we have introduced, in the line of [3] and [27]. Note that the Donsker delta function has been used for giving explicit representation formulae for the hedging portfolio in some market models driven by Brownian motion or pure jump Lévy processes, see [3] and [9].

Definition 3.1. For a given random variable $X \in (\mathcal{S})_{-1}$ the Donsker delta function of X is a continuous function $\delta_\bullet(X) : \mathbb{R} \to (\mathcal{S})_{-1}$ such that

$$\int_{\mathbb{R}} h(u)\, \delta_u(X)\, du = h(X)$$

for all Borel real functions h on $\mathbb{R}$ for which the integral is well-defined in $(\mathcal{S})_{-1}$. Following [27] we consider a measure P_2 in $P = P_1 \otimes P_2$ (see (4)) which satisfies the condition: there exists $\epsilon \in (0,1)$ such that

$$\lim_{|u|\to\infty} |u|^{-(1+\epsilon)} \operatorname{Re}\left\{ \int_{\mathbb{R}_0} \left(e^{iuz} - 1 - iuz\right) \nu(dz) \right\} = \infty\,. \tag{26}$$

Using Theorem 3.1 we can obtain an explicit formula for the Donsker delta function of $\xi(t)$:

Theorem 3.2. *Assume that (26) holds. Then the Donsker delta function $\delta_u(\xi^y(t))$, $u \in \mathbb{R}$, of $\xi^y(t)$, $t \in [0,T]$, exists in $(\mathcal{S})^*$ and is given by*

$$\begin{aligned}\delta_u(\xi^y(t)) = \frac{1}{2\pi} \int_{\mathbb{R}} \exp^\diamond \Bigg[\int_0^t \int_{\mathbb{R}} \left(e^{i\lambda\gamma(s,z)} - 1\right) \tilde{N}(ds,dz) + \int_0^t i\lambda\, \beta(s)\, dB(s) \\ + \int_0^t \Bigg\{ \int_{\mathbb{R}} \left(e^{i\lambda\gamma(s,z)} - 1 - i\lambda\gamma(s,z)\right) \nu(dz) \\ + i\lambda\alpha(s) - \tfrac{1}{2}\lambda^2\beta^2(s) \Bigg\} ds + i\lambda y - i\lambda u \Bigg] d\lambda.\end{aligned} \tag{27}$$

Proof (sketch). Formally this follows from (22) by using the Fubini theorem in $(\mathcal{S})^*$, as follows. By (22) we have

$$\begin{aligned} g(\xi^y(t)) &= \int_{\mathbb{R}} \hat{g}(\lambda) \exp^\diamond(X^y_\lambda(t))\, d\lambda \\ &= \frac{1}{2\pi} \int_{\mathbb{R}} \left(\int_{\mathbb{R}} e^{-i\lambda u} g(u)\, du \right) \exp^\diamond\left(X^y_\lambda(t)\right) d\lambda \\ &= \frac{1}{2\pi} \int_{\mathbb{R}} g(u) \int_{\mathbb{R}} \exp^\diamond\left(-i\lambda u + X^y_\lambda(t)\right) d\lambda\, du. \end{aligned}$$

For justification and more details we refer to the proof in [27, Theorem 3.1.4]. □

4. Application to sensitivity with respect to the starting point

Let $X(t) = X^x(t)$, $t \in [0,T]$, be a jump diffusion of the form

$$\begin{cases} dX(t) = X(t^-)\left[\mu(t)dt + \sigma(t)dB(t) + \int_{\mathbb{R}} \theta(t,z)\tilde{N}(dt,dz)\right] \\ X(0) = x > 0 \end{cases} \tag{28}$$

where $\mu(t), \sigma(t)$ and $\theta(t,z)$, $t \in [0,T]$, $z \in \mathbb{R}$, are deterministic, $\theta(t,z) > -1$ for a.a. t, z and

$$\int_0^T \left\{ |\mu(t)| + \sigma^2(t) + \int_{\mathbb{R}} \theta^2(t,z)\,\nu(dz) \right\} dt < \infty$$

— cf. (3). By the Itô formula for Lévy processes (see e.g. [30, Theorem 1.14], the solution of this equation is

$$\begin{aligned} X^x(t) = x \exp\Bigg[\int_0^t \left\{ \mu(s) - \tfrac{1}{2}\sigma^2(s) + \int_{\mathbb{R}} \big(\ln(1+\theta(s,z)) - \theta(s,z)\big)\,\nu(dz) \right\} ds \\ + \int_0^t \sigma(s)\,dB(s) + \int_0^t \int_{\mathbb{R}} \ln(1+\theta(s,z))\,\tilde{N}(ds,dz) \Bigg] \\ = \exp(\xi^y(t)), \end{aligned} \tag{29}$$

where $d\xi^y(t) = \alpha(t)\,dt + \beta(t)\,dB(t) + \int_{\mathbb{R}} \gamma(t,z)\,\tilde{N}(dt,dz)$, with

$$\begin{aligned} \alpha(t) &= \mu(t) - \tfrac{1}{2}\sigma^2(t) + \int_{\mathbb{R}} \big(\ln(1+\theta(t,z)) - \theta(t,z)\big)\,\nu(dz), \\ \beta(t) &= \sigma(t), \quad \gamma(t,z) = \ln(1+\theta(t,z)) \quad \text{and} \quad y = \ln x \end{aligned} \tag{30}$$

— cf. (16). Therefore, if $h : \mathbb{R} \to \mathbb{R}$ then

$$E[h(X^x(T))] = E[h(\exp(\xi^y(T)))] = E[g(\xi^y(T))],$$

where

$$g(u) := h(\exp(u)), \quad u \in \mathbb{R}.$$

If this g satisfies the conditions of Theorem 3.1 then

$$\begin{aligned} \frac{d}{dx} E[h(X^x(T))] &= \frac{d}{dx}[Eg(\xi^{\ln x}(T))] = \frac{d}{dx} \int_{\mathbb{R}} \hat{g}(\lambda) \exp(i\lambda \ln x + G_\lambda(T))\,d\lambda \\ &= \int_{\mathbb{R}} \hat{g}(\lambda)\,\frac{i\lambda}{x}\,\exp(i\lambda \ln x + G_\lambda(T))\,d\lambda, \end{aligned}$$

where $G_\lambda(T)$ is given by (25). We have proved

Theorem 4.1. *Suppose $h : \mathbb{R} \to \mathbb{R}$ is such that $g(u) := h(\exp(u))$, $u \in \mathbb{R}$, satisfies the conditions of Theorem 3.1 and that*

$$\int_{\mathbb{R}} \big|\hat{g}(\lambda)\,\lambda\,\exp\{\operatorname{Re} G_\lambda(T)\}\big|\,d\lambda < \infty\,.$$

Then

$$\frac{d}{dx}E[h(X^x(T))] = \int_{\mathbb{R}} \hat{g}(\lambda)\,\frac{i\lambda}{x}\,\exp\big(i\lambda\ln x + G_\lambda(T)\big)\,d\lambda\,. \tag{31}$$

Example 4.1. Choose $h(u) = \chi_{[H,K]}(u)$, $u \in \mathbb{R}$ $(H, K > 0)$. Then $h(X^x(T))$ may be regarded as the payoff of a digital option on a stock with price $X^x(T)$. In this case

$$g(u) = \chi_{[H,K]}(e^u), \quad u \in \mathbb{R},$$

and

$$\begin{aligned} 2\pi\,\hat{g}(\lambda) &= \int_{\mathbb{R}} e^{-i\lambda u} g(u)\,du = \int_{\mathbb{R}} e^{-i\lambda u}\chi_{[H,K]}(e^u)\,du \\ &= \int_{\ln H}^{\ln K} e^{-i\lambda u}\,du = \frac{H^{-i\lambda} - K^{-i\lambda}}{i\lambda}\,. \end{aligned}$$

Therefore

$$\frac{d}{dx}E\big[\chi_{[H,K]}(X^x(T))\big] = \int_{\mathbb{R}} \frac{H^{-i\lambda} - K^{-i\lambda}}{x}\exp\big(i\lambda\ln x + G_\lambda(T)\big)\,d\lambda\,, \tag{32}$$

provided that the integral converges.

A sufficient condition for this is that, for some $\delta > 0$,

$$\lambda^2 \int_0^T \Big\{\beta^2(s) + \int_{\mathbb{R}} \big(1 - \cos(\lambda\gamma(s,z))\big)\,\nu(dz)\Big\}\,ds \geq \delta\lambda^2 \quad \text{for all } \lambda \in \mathbb{R},$$

which is a weak form of non-degeneracy of the equation (28). Thus, in spite of the fact that h is not even continuous, (31) is a computationally efficient formula for $\frac{d}{dx}E^x[h(X^x(T))]$.

References

1. S. Albeverio, Y. G. Kondratiev, L. Streit, "How to generalize white noise analysis to non-Gaussian spaces", in Ph. Blanchard et al. (eds), *Dynamics of Complex and Irregular Systems*, World Scientific, 1993.
2. K. Aase, B. Øksendal, J. Ubøe, "Using the Donsker delta function to compute hedging strategies", *Potential Analysis* **14**, 351–374 (2001).

3. K. Aase, B. Øksendal, N. Privault, J. Ubøe, "White noise generalizations of the Clark-Hausmann-Ocone theorem with application to mathematical finance", *Finance Stoch.* **4**, 465–496 (2000).
4. M.-P. Bavouzet, M. Massaoud, "Computation of Greeks using Malliavin's calculus in jump-type market models", Report 5482, INRIA, Rocquencourt, France, 2005.
5. E. Benhamou, "Optimal Malliavin weighting function for the computation of the Greeks", Conference on Applications of Malliavin Calculus in Finance (Rocquencourt, 2001). *Math. Finance* **13**, 37–53 (2003).
6. Yu. M. Berezansky, Yu. G. Kontratiev, *Spectral Methods in Infinite-Dimensional Analysis*, Kluwer Academic Publishers, 1995.
7. J. Bertoin, *Lévy processes*, Cambridge University Press, 1996.
8. M. H. A. Davis, M. P. Johansson, "Malliavin Monte Carlo Greeks for jump diffusions", *Stochastic Process. Appl.* **116**, 101–129 (2006).
9. G. Di Nunno, B. Øksendal, "The Donsker delta function, a representation formula for functionals of a Lévy process and application to hedging in incomplete markets", Preprint Series in Pure Math. 11, Dept. of Mathematics, University of Oslo (2004). To appear in *Sem. Congres. Ac. Sci.*
10. G. Di Nunno, B. Øksendal, F. Proske, "White noise analysis for Lévy processes", *Journal of Functional Analysis* **206**, 109–148 (2004).
11. Y. El-Khatib, N. Privault, "Computations of Greeks in a market with jumps via the Malliavin calculus", *Finance Stoch.* **8**, 161–179 (2004).
12. E. Fournié, J.-M. Lasry, J. Lebuchoux, P.-L. Lions, "Applications of Malliavin calculus to Monte-Carlo methods in finance. II", *Finance Stoch.* **5**, 201–236 (2001).
13. E. Fournié, J.-M. Lasry, J. Lebuchoux, P.-L. Lions, N. Touzi, "Applications of Malliavin calculus to Monte-Carlo methods in finance." *Finance Stoch.* **3**, 391–412 (1999).
14. P. W. Glynn, "Optimization of stochastic systems via simulation", in *Proceedings of the 1989 Winter simulation Conference*, San Diego: Society for Computer Simulation, 1989, pp. 90–105.
15. E. Gobet, R. Munos, "Sensitivity analysis using Itô-Malliavin calculus and martingales, and application to stochastic optimal control", *SIAM J. Control Optim.* **43**, 1676–1713 (2005).
16. P. Glasserman, D. D. Yao, "Some guidelines and guarantees for common random numbers", *Manag. Sci.* **38**, 884–908 (1992).
17. T. Hida, "White noise analysis and its applications", in L.H.Y. Chen (ed.), *Proc. Int. Mathematical Conf.*, North-Holland, Amsterdam, 1982, pp. 43–48.
18. T. Hida, H.-H. Kuo, J. Potthoff, L. Streit, *White Noise*, Kluwer, Dordrecht, 1993.
19. H. Holden, B. Øksendal, J. Ubøe, T.-S. Zhang, *Stochastic Partial Differential Equations — A Modeling, White Noise Functional Approach.* Birkhäuser, Boston, 1996.
20. K. Itô, "Spectral type of the shift transformation of differential processes with stationary increments", *Trans. Am. Math. Soc.* **81**, 253–263 (1956).

21. Yu. G. Kondratiev, *Generalized Functions in Problems in Infinite Dimensional Analysis*, Ph.D. Thesis, University of Kiev, 1978.
22. Y. Kondratiev, J. L. Da Silva, L. Streit, "Generalized Appell systems", *Methods Funct. Anal. Topology* **3**, 28–61 (1997).
23. Y. Kondratiev, J. L. Da Silva, L. Streit, G. Us, "Analysis on Poisson and gamma spaces", *Inf. Dim. Anal. Quant. Prob. Rel. Topics* **1**(1), 91–117 (1998).
24. A. Kohatsu-Higa, M. Montero, "Malliavin calculus in finance", *Handbook of computational and numerical methods in finance*, Birkhäuser, 2004, pp. 111–174.
25. H. H. Kuo, *White Noise Distribution Theory, Prob. and Stoch. Series*, Boca Raton, FL, CRC Press, 1996.
26. P. Malliavin, A. Thalmaier, *Stochastic calculus of variations in mathematical finance*, Springer Finance, 2006.
27. S. Mataramvura, B. Øksendal, F. Proske, "The Donsker delta functin of a Lévy process with application to chaos expansion of local time", *Ann. Inst. H. Poincaré Probab. Statist.* **40**, 553–567 (2004).
28. N. Obata, *White Noise Calculus and Fock Space*, LNM, 1577, Springer-Verlag, Berlin, 1994.
29. B. Øksendal, F. Proske, "White noise of Poisson random measures", *Potential Analysis* **21**, 375–403 (2004).
30. B. Øksendal, A. Sulem, *Applied Stochastic Control of Jump Diffusions*, second edition, Springer, 2006.
31. N. Privault, "Splitting of Poisson noise and Lévy processes on real Lie algebras", *Infin. Dimen. Anal. Quantum Probab. Relat. Top.* **5**, 21–40 (2002).
32. N. Privault, X. Wei, "A Malliavin calculus approach to sensitivity analysis in insurance", *Insurance Math. Econom.* **35**, 679–690 (2004).
33. K. Sato, *Lévy Processes and Infinitely Divisible Distributions*, Cambridge University Press, 1999.

GIULIA DI NUNNO
Centre of Mathematics for Applications (CMA),
Dept. of Mathematics, University of Oslo,
P.O. Box 1053 Blindern, N–0316 Oslo, Norway
giulian@math.uio.no

BERNT ØKSENDAL
Centre of Mathematics for Applications (CMA),
Dept. of Mathematics, University of Oslo,
P.O. Box 1053 Blindern, N–0316 Oslo, Norway
and
Norwegian School of Economics and Business Administration,
Helleveien 30, N–5045 Bergen, Norway
oksendal@math.uio.no

CREATION AND ANNIHILATION OPERATORS ON LOCALLY COMPACT SPACES

WILHELM VON WALDENFELS (*Universität Heidelberg*)

Denote by ε_x the point measure in the point x of a locally compact space X. Define the annihilation operator $a(\varepsilon_x)$ and the creation operator $a^+(\varepsilon_.)(dx)$. We establish the usual algebraic relations and prove a generalization of Wick's theorem. If a privileged measure on X is introduced, a duality theory can be established similar to the habitual one. We derive a generalization of a convolution formula due to Meyer.

1. Introduction

A basic formula of quantum probability is

$$[a_s, a_t^+] = \delta(s-t),$$

the commutator of the annihilation operator in the point $s \in \mathbb{R}$ with the creation operator in the point $t \in \mathbb{R}$. As the appearance of Dirac's δ-function indicates, a_t and a_t^+ are highly singular. Hudson and Parasarathy [11] choose a way out in working with the differentials

$$dA_t = \int_t^{t+dt} a_s\, ds, \quad dA_t^+ = \int_t^{t+dt} a_s^+\, ds, \quad d\Lambda_t = \int_t^{t+dt} a_s^+ a_s\, ds.$$

Another solution is to use white noise analysis as proposed by Obata [10]. As the physicists use the calculus based on the commutation relation naively and with big success, Accardi, Lu and Volovich [1] attempted to put the physicists' calculus on a solid base using the theory of Schwartz distributions.

An old tool of physicists is to represent operators in a form of a power series

$$T(f) = \sum_{l,m,n=0}^{\infty} \frac{1}{l!\, m!\, n!} \int f(s_1, \ldots, s_l; t_1, \ldots, t_m; u_1, \ldots, u_n)$$
$$a_{s_1}^+ \cdots a_{s_l}^+ a_{t_1}^+ \cdots a_{t_m}^+ a_{t_1} \cdots a_{t_m} a_{u_1} \cdots a_{u_n}$$
$$ds_1 \cdots ds_l\, dt_1 \cdots dt_m\, du_1 \cdots du_n\,.$$

The idea introduced by Maassen [8] into quantum probability was to establish a formula

$$T(f)T(g) = T(f * g),$$

and to work with the convolution $f, g \mapsto f * g$ instead with the operators a_t, a_t^+. Maassen did so with operators not including the number operator, i.e., where the term $a_{t_1}^+ \cdots a_{t_m}^+ a_{t_1} \cdots a_{t_m}$ does not arise. Meyer [9] succeeded to give a formula for the convolution including the number operator. The analysis using convolution becomes a lot easier, but the computations get very difficult, especially in the case with the number operator included, if one has to do with the complicated formula due to Meyer [9, p. 92].

In a previous paper the author generalized Maassen's approach to locally compact spaces in order to use measures in the place of test functions [15]. But it seemed to be too cumbersome to include Meyer's theory. So we turn to the ideas of Accardi, Lu, Volovich, and try to imitate in this paper the physicists' approach by directly dealing with the operators a_t, a_t^+. We do not use Schwartz distributions, but deal with measure theory in the sense of Bourbaki's integration theory, especially his chapter about integration of measures [4, 5].

Instead of $\mathbb{R}$ and the usual Fock space we work with a locally compact space X and the space $\mathfrak{X}$ of finite sequences of elements of X. We have no privileged measure on X like the Lebesgue measure in the usual theory. If ε_x is the point measure in the point $x \in X$, then a_t and a_t^+ are replaced by

$$\begin{aligned} a(\varepsilon_x):\ & (a(\varepsilon_x)f)(x_1, \ldots, x_n) = f(x, x_1, \ldots, x_n) \\ a^+(\varepsilon_.)(dx):\ & (a^+(\varepsilon_.)f)(x_1, \ldots, x_n; dx) \\ & = \varepsilon_{x_1}(dx) f(x_2, \ldots, x_n) + \cdots + \varepsilon_{x_n}(dx) f(x_1, \ldots, x_{n-1}). \end{aligned}$$

In section 2 we introduce the basic notations. In section 3 we study admissible monomials, these are monomials in $a(\varepsilon_x), a^+(\varepsilon_.)(dx)$, where any variable may arise basically once with the exception that monomials of the form

$$* \cdots * a^+(\varepsilon_.)(dx) * \cdots * a(\varepsilon_x) * \cdots *$$

are allowed. In the contrary monomials of the form

$$* \cdots a(\varepsilon_x) * \cdots * a^+(\varepsilon_.)(dx) * \cdots *$$

are forbidden.

In section 4 we prove a generalization of Wick's theorem. In section 5 we introduce a distinguished measure replacing Lebesgue measure in the usual theory and treat duality. In order to check our theory we derive in section 6 the generalization of Meyer's convolution formula.

2. Preliminaries.

Let X be a set. We denote by $\mathfrak{W}(X)$ the set of all finite sequences of elements of X or words formed by elements of X and write for short $\mathfrak{X} = \mathfrak{W}(X)$ and we have

$$\mathfrak{X} = \emptyset + X + X^2 + \cdots$$

We use the plus sign for denoting disjoint union of sets. We denote by $\mathfrak{S}(X)$ the set of finite subsets of X and by $\mathfrak{M}(X)$ the set of finite multisets of X. A multiset is a pair $m = (S, r)$, where S is a set and r is a function $r : S \to \mathbb{N} = \{0, 1, \ldots\}$. The cardinality of m is $\#m = \sum_{s \in S} r(s)$. The multiset is finite if its cardinality is finite. We may write multisets in the form $m = \{a_1, \ldots, a_n\}$, where not all the a_i must be different. So, e.g., $\{2, 1, 1, 2, 3\}$ is a multiset of the set $\{1, 2, 3, 4\}$ with $r(1) = r(2) = 2$, $r(3) = 1$, $r(4) = 0$ and $\{1, 1, 2, 2, 3\}$ denotes the same multiset.

If $w = (x_1, \cdots, x_n)$ is a word and σ is a permutation, then

$$\sigma w = (x_{\sigma^{-1}(1)}, \ldots, x_{\sigma^{-1}(n)}).$$

The word w defines a multiset

$$m_w = \{x_1, \ldots, x_n\}.$$

If w' is another word, then $m_{w'} = m_w$ iff there exists a permutation changing w into w'.

An n-chain is a totally ordered set of n elements

$$\underline{a} = (a_1, \ldots, a_n).$$

We denote by $x_{\underline{a}}$ the word

$$x_{\underline{a}} = (x_{a_1}, \ldots, x_{a_n}).$$

If α is the underlying set of $\underline{a}$, we denote by

$$x_\alpha = \{x_{a_1}, \ldots, x_{a_n}\}$$

the corresponding multiset, as it does not depend on the order of $\underline{a}$.

A function f on $\mathfrak{X}$ is called symmetric, if $f(w) = f(\sigma w)$ for all permutations of w. If α is an n-set, i.e., a set with n elements, then $f(x_\alpha)$ is well defined.

Assume now that X is a locally compact space denumerable at infinity, provide X^n with the product topology and $\mathfrak{X} = \mathfrak{W}(X)$ with that topology where the X^n are as well open and closed and where the restriction to X^n coincides with the topology of X^n. Then $\mathfrak{X}$ is locally compact as well, its compact sets are contained in a finite union of the X^n and their intersections with the X^n are compact.

If S is a locally compact space, denote by $\mathcal{K}(S)$ the space of complex valued continuous functions on S with compact support and by $\mathcal{M}(S)$ the space of complex measures on S. If μ is a complex measure on $\mathfrak{X}$, we write

$$\mu = \mu_0 + \frac{1}{1!}\mu_1 + \frac{1}{2!}\mu_2 + \cdots$$

where $\mu_n/n!$ is the restriction of μ to X^n. We denote by Ψ the measure given by

$$\Psi(f) = f(\emptyset). \tag{1}$$

Then μ_0 is a multiple of Ψ. If $\underline{a}$ is an n-chain, we denote

$$\mu(dx_{\underline{a}}) = \mu_n(dx_{a_1}, \ldots, dx_{a_n}).$$

If μ is a symmetric measure, then $\mu(dx_\alpha)$, where α is an n-set is well defined.

One has

$$\int \mu(dw) f(w) = \mu_0 f(\emptyset) + \sum_{n=1}^{\infty} \frac{1}{n!} \int \mu_n(dx_1, \ldots, dx_n)\, f(x_1, \ldots, x_n)$$

A *hierarchy* is a family of finite index sets

$$A = \{\alpha_0, \alpha_1, \alpha_2, \ldots\},$$

such that $\#\alpha_n = n$. We write

$$\begin{aligned}\int \mu(dw)\, f(w) &= \int_{\alpha\in A} \mu(dx_\alpha)\, f(x_\alpha)\\ &= \mu_0 f(\emptyset) + \sum_{n=1}^{\infty} \frac{1}{n!} \int \mu_n(dx_{\alpha_n})\, f(x_{\alpha_n})\end{aligned}$$

or simply

$$\int_\alpha \mu(dx_\alpha)\, f(x_\alpha)$$

or

$$\int_\alpha \mu(\alpha)\, f(\alpha)$$

if the variable x is clear. Recall lemma 2.1 [15].

Lemma 2.1 (Sum-integral lemma for measures). *Let be given a bounded measure*

$$\mu(dw_1, \ldots, dw_k)$$

on

$$\mathfrak{X}^k = \sum_{n_1,\ldots,n_k} X^{n_1} \times \cdots X^{n_k},$$

symmetric in each of the variables w_i *and write*

$$\mu = \sum \frac{1}{n_1! \cdots n_k!} \mu_{n_1,\ldots,n_k}$$

where

$$\frac{1}{n_1! \cdots n_k!} \mu_{n_1,\ldots,n_k}$$

is the restriction of μ *to* $X^{n_1} \times \cdots \times X^{n_k}$. *Then*

$$\int \cdots \int \mu(dw_1, \ldots, dw_k) = \int \nu(dw),$$

where ν *is a bounded measure on* $\mathfrak{X}$ *and*

$$\nu = \sum \frac{1}{n!} \nu_n.$$

Here $\nu_n/n!$ *is the restriction of* ν *to* X^n *and*

$$\nu_n(dx_1, \ldots, dx_n) = \sum_{\beta_1+\cdots+\beta_k=[1,n]} \mu_{\#\beta_1,\ldots,\#\beta_k}(dx_{\beta_1}, \ldots, dx_{\beta_k}),$$

where $\beta_1, \ldots, \beta_k$ *are disjoint sets.*

Using hierarchies $A_1, \ldots, A_k, B$ *we may write*

$$\int_{\alpha_1 \in A_1} \cdots \int_{\alpha_k \in A_k} \mu(dx_{\alpha_1}, \ldots, dx_{\alpha_k}) = \int_{\beta \in B} \nu(dx_\beta)$$

with

$$\nu(dx_\beta) = \sum_{\beta_1+\cdots+\beta_k=\beta} \mu(dx_{\beta_1}, \ldots, dx_{\beta_k})$$

$$\mu(dx_{\beta_1}, \ldots, dx_{\beta_k}) = \mu_{\#\beta_1,\ldots,\#\beta_k}(dx_{\beta_1}, \ldots, dx_{\beta_k}).$$

We denote by

$$\mathcal{K} = \mathcal{K}_s(\mathfrak{X}) \tag{2}$$

the space of continuous, *symmetric* functions with compact support in $\mathfrak{X}$.

We define creation and annihilation operators for symmetric functions and measures on $\mathfrak{X}$. Assume a function $\varphi \in \mathcal{K}(X)$, a function $f \in \mathcal{K}$, a measure $\nu \in \mathcal{M}(X)$, a symmetric measure $\mu \in \mathcal{M}(\mathfrak{X})$. We define

$$(a(\nu)f)(x_1, \ldots, x_n) = \int \nu(dx_0)\, f(x_0, x_1, \ldots, x_n)$$

or, in another notation, where $\alpha + c = \alpha + \{c\}$ means that the point c is added to the set α and similar using $\alpha \setminus c = \alpha \setminus \{c\}$

$$(a(\nu)f)(x_\alpha) = \int \nu(dx_c)\, f(x_{\alpha+c}), \tag{3}$$

$$(a^+(\varphi)f)(x_\alpha) = \sum_{c\in\alpha} \varphi(x_c)\, f(x_{\alpha\setminus c}), \tag{4}$$

$$(a^+(\nu)\mu)(dx_\alpha) = \sum_{c\in\alpha} \nu(dx_c)\, \mu(dx_{\alpha\setminus c}), \tag{5}$$

$$(a(\varphi)\mu)(dx_\alpha) = \int \varphi(x_c)\, \mu(dx_{\alpha+c}). \tag{6}$$

If Φ is the function

$$\Phi(\emptyset) = 1, \;\; \Phi(x_\alpha) = 0 \text{ for } \alpha \neq \emptyset, \tag{7}$$

then

$$a(\nu)\,\Phi = 0.$$

Similarly if Ψ is the measure defined in (1), then

$$a(\varphi)\Psi = 0.$$

One finds the commutation relations

$$\left[a(\nu), a^+(\varphi)\right] = \int \nu(dx)\, \varphi(x), \tag{8}$$

$$\left[a(\varphi), a^+(\nu)\right] = \int \nu(dx)\, \varphi(x), \tag{9}$$

and obtains

$$\int (a^+(\nu)\mu)(dw)\, f(w) = \int \mu(dw)\, (a(\nu)f)(w), \tag{10}$$

$$\int (a(\varphi)\mu)(dw)\, f(w) = \int \mu(dw)\, (a^+(\varphi)f)(w). \tag{11}$$

We define the exponential measures and functions

$$e(\varphi) = \Phi + \varphi + \varphi^{\otimes 2} + \cdots = e^{a^+(\varphi)}\Phi, \tag{12}$$

$$e(\nu) = \Psi + \nu + \frac{1}{2!}\nu^{\otimes 2} + \cdots = e^{a^+(\nu)}\Psi, \tag{13}$$

as $a^+(\nu)^n \Psi / n! = \nu^{\otimes n}$. So for $\alpha = \{a_1, \ldots, a_n\}$

$$e(\varphi)(x_\alpha) = \varphi(x_{a_1}) \cdots \varphi(x_{a_n}), \tag{14}$$

$$e(\nu)(dx_\alpha) = \nu(dx_{a_1}) \cdots \nu(dx_{a_n}). \tag{15}$$

If S is a locally compact space and μ a measure on S and f a Borel function, we define the product $f\mu$ by

$$\int (f\mu)(ds)\, \varphi(s) = \int \mu(ds)\, f(s)\, \varphi(s)$$

for $\varphi \in \mathcal{K}(S)$ and write

$$(f\mu)(ds) = (\mu f)(ds) = f(s)\, \mu(ds).$$

Let S and T be locally compact spaces. We consider a function $f : S \to \mathcal{M}(T)$. It can be considered as a function

$$f : S \times \mathcal{K}(T) \to \mathbb{C}$$

and we write it

$$f = f(s, dt).$$

We extend the notion of the creation operator to functions $f = f(x, dy) : X \to \mathcal{M}(X)$ and define for $g \in \mathcal{K}_s(\mathfrak{X})$

$$(a^+(f)g)(x_\alpha, dy) = \sum_{c \in \alpha} f(x_c, dy)\, g(x_{\alpha \setminus c}).$$

We consider the function

$$\varepsilon : x \in X \mapsto \varepsilon_x \in \mathcal{M}(X)$$

$$\int \varepsilon_x(dy)\, \varphi(y) = \varphi(x),$$

so ε_x is the point measure in the point x, and

$$(a^+(\varepsilon)g)(x_\alpha, dy) = \sum_{c \in \alpha} \varepsilon_{x_c}(dy)\, g(x_{\alpha \setminus c}).$$

We may consider $a^+(\varepsilon)$ as an operator valued measure and write

$$a^+(\varepsilon) = a^+(\varepsilon)(dy). \tag{16}$$

We obtain the commutation relations

$$[a(\varepsilon_x), a(\varepsilon_y)] = 0 \tag{17}$$

$$\left[a^+(\varepsilon)(dx), a^+(\varepsilon)(dy)\right] = 0 \tag{18}$$

$$\left[a(\varepsilon_x), a^+(\varepsilon)(dy)\right] = \varepsilon_x(dy). \tag{19}$$

We extend this notion to a Borel function $g : y \in X \mapsto g_y \in \mathcal{K}_s(\mathfrak{X})$ and write

$$(a^+(\varepsilon)g_y)(x_\alpha, dy) = \sum_{c\in\alpha} \varepsilon_{x_c}(dy)\, g_y(x_{\alpha\setminus c}).$$

Here appears the product of the measure $\varepsilon_{x_c}(dy)$ with the function g_y.

A special case arises if $g_y = a(\varepsilon_y)f$. Then

$$(a^+(\varepsilon)a(\varepsilon_y)f)(x_\alpha, dy) = \sum_{c\in\alpha} \varepsilon_{x_c}(dy)\, g(x_\alpha).$$

So

$$\mathfrak{n}(dy) = a^+(\varepsilon)(dy)\, a(\varepsilon_y) \tag{20}$$

is the operator analogous to the number operator $a_t^+ a_t$.

3. Admissible monomials

Recall the definition of the space $\mathcal{K}$ of all continuous symmetric functions with compact support in $\mathfrak{X}$. Assume π and ϱ two disjoint index sets and define $\mathcal{K}_{\pi,\varrho}$ as the space of all functions

$$f : \mathfrak{X} \times X^\varrho \to \mathcal{M}(X^\pi)$$

with the following property:

$$(w, \varphi, y) \in \mathfrak{X} \times \mathcal{K}(X^\pi) \times X^\varrho \to \int_{X^\pi} f(w, dx, y)\, \varphi(x)$$

is symmetric in w, continuous and for fixed φ a continuous function with compact support in $\mathfrak{X} \times X^\varrho$.

For $c \notin \pi + \varrho$ define

$$a_c : \mathcal{K}_{\pi,\varrho} \to \mathcal{K}_{\pi,\varrho+c}$$

$$(a_c f)(x_\alpha, (dx_p)_{p\in\pi}, (x_r)_{r\in\varrho+c}) = f(x_{\alpha+c}, (dx_p)_{p\in\pi}, (x_r)_{r\in\varrho})$$

and for $c \notin \pi$

$$a_c^+ : \mathcal{K}_{\pi,\varrho} \to \mathcal{K}_{\pi+c,\varrho\backslash c}$$
$$(a_c^+ f)(x_\alpha, (dx_p)_{p\in\pi+c}, (x_r)_{r\in\varrho\backslash c}) = \sum_{b\in\alpha} \varepsilon_{x_b}(dx_c)\, f(x_{\alpha\backslash b}, (dx_p)_{p\in\pi}, (x_r)_{r\in\varrho}).$$

Here

$$\varrho \setminus c = \varrho \setminus (\varrho \cap \{c\}) = \begin{cases} \varrho & \text{for } c \notin \varrho, \\ \varrho \setminus c & \text{for } c \in \varrho. \end{cases}$$

If $c \notin \varrho$ the right-hand side yields

$$\sum_{b\in\alpha} \varepsilon_{x_b}(dx_c)\, f(x_{\alpha\backslash b}, (dx_p)_{p\in\pi}, (x_r)_{r\in\varrho})$$

and for $c \in \varrho$ it yields

$$\sum_{b\in\alpha} \varepsilon_{x_b}(dx_c)\, f(x_{\alpha\backslash b}, (dx_p)_{p\in\pi}, (x_c, (x_r)_{r\in\varrho\backslash c})).$$

One calculates easily

Lemma 3.1. *Assume $c \neq c'$ and $f \in \mathcal{K}_{\pi,\varrho}$. If $c, c' \notin \pi + \varrho$ then*

$$a_c a_{c'} f = a_{c'} a_c f,$$

if $c, c' \notin \pi$ then

$$a_c^+ a_{c'}^+ f = a_{c'}^+ a_c^+ f,$$

and if $c \notin \pi + \varrho$ and $c' \notin \pi$ then

$$a_c a_{c'}^+ f = \varepsilon_{x_c}(dx_{c'}) f + a_{c'}^+ a_c f.$$

Definition 3.1. A sequence

$$W = (a_{c_n}^{\theta_n}, \ldots, a_{c_1}^{\theta_1})$$

with $c_1, \ldots, c_n$ indices and $\theta_i = \pm 1$ and

$$a_c^\theta = \begin{cases} a_c^+ & \text{for } \theta = +1 \\ a_c & \text{for } \theta = -1 \end{cases}$$

is called *admissible* if

$$i > j \implies \{c_i \neq c_j \text{ or } \{c_i = c_j \text{ and } \theta_i = 1, \theta_j = -1\}\}.$$

So W is admissible if it contains only pairs (not necessarily being neighbors) of the form $(a_c^\theta, a_{c'}^{\theta'})$ with $c \neq c'$ or (a_c^+, a_c) and no pairs of the form (a_c, a_c), (a_c^+, a_c^+) or (a_c, a_c^+).

Definition 3.2. If W is an admissible sequence, define

$$\begin{aligned} \omega(W) &= \{c_1, \ldots, c_n\} \\ \omega_+(W) &= \{c_i, 1 \leq i \leq n : \theta_i = +1\} \\ \omega_-(W) &= \{c_i, 1 \leq i \leq n : \theta_i = -1\}. \end{aligned}$$

Lemma 3.2. *If W is admissible and $W_1 \subset W$, then W_1 is admissible. If W is admissible, then*

$$\begin{aligned} (a_c^+, W) \text{ is admissible} &\iff c \notin \omega_+(W) \\ (a_c, W) \text{ is admissible} &\iff c \notin \omega(W) \\ (W, a_c^+) \text{ is admissible} &\iff c \notin \omega(W) \\ (W, a_c) \text{ is admissible} &\iff c \notin \omega_-(W). \end{aligned}$$

If $W = (W_2, W_1)$ is admissible, then

$$\begin{aligned} (W_2, a_c^+, W_1) \text{ is admissible} &\iff c \notin \omega(W_2) \cup \omega_+(W_1) \\ (W_2, a_c, W_1) \text{ is admissible} &\iff c \notin \omega_-(W_2) \cup \omega(W_1). \end{aligned}$$

Proposition 3.1. *Assume*

$$W = (a_{c_n}^{\theta_n}, \ldots, a_{c_1}^{\theta_1})$$

to be an admissible sequence. Assume disjoint index sets π and ϱ and

$$\begin{aligned} \omega_+(W) \cap \pi &= \emptyset \\ \omega_-(W) \cap (\pi + \varrho) &= \emptyset. \end{aligned}$$

Define for $k = 1, \ldots, n$

$$W_k = (a_{c_k}^{\theta_k}, \cdots, a_{c_1}^{\theta_1}).$$

Set $\pi_0 = \pi, \varrho_0 = \varrho$ and

$$\begin{aligned} \pi_k &= \pi + \omega_+(W_k), \\ \varrho_k &= \varrho \setminus (\omega_+(W_k) \setminus \omega_-(W_k)) + \omega_-(W_k) \setminus \omega_+(W_k), \end{aligned}$$

where for sets α, β

$$\alpha \setminus \beta = \alpha \setminus (\alpha \cap \beta).$$

Then

$$a_{c_k}^{\theta_k} : \mathcal{K}_{\pi_{k-1}, \varrho_{k-1}} \to \mathcal{K}_{\pi_k, \varrho_k}$$

and the iterated application

$$M = a_{c_n}^{\theta_n} \cdots a_{c_1}^{\theta_1} : \mathcal{K}_{\pi, \varrho} \to \mathcal{K}_{\pi', \varrho'}$$

with

$$\begin{aligned} \pi' &= \pi + \omega_+(W), \\ \varrho' &= \varrho \setminus (\omega_+(W) \setminus \omega_-(W)) + \omega_-(W) \setminus \omega_+(W). \end{aligned}$$

Definition 3.3. If

$$W = (a_{c_n}^{\theta_n}, \ldots, a_{c_1}^{\theta_1})$$

is an admissible sequence we call

$$M = a_{c_n}^{\theta_n} \cdots a_{c_1}^{\theta_1}$$

an admissible monomial.

Proof. We prove by induction. The case $k = 1$ is trivial. Put $\omega_k = \omega(W_k)$, etc.. Assume

$$\begin{aligned} \pi_{k-1} &= \pi + \omega_{+,k-1}, \\ \varrho_{k-1} &= \varrho \setminus (\omega_{+,k-1} \setminus \omega_{-,k-1}) + \omega_{-,k-1} \setminus \omega_{+,k-1}\,. \end{aligned}$$

Assume $\theta_k = +1$. In order that $a_{c_k}^+$ is defined, $c_k \notin \pi_{k-1}$. But $c_k \notin \pi$ by assumption and $c_k \notin \omega_{+,k-1}$, as W_k is admissible. So

$$a_{c_k}^+ : \mathcal{K}_{\pi_{k-1}, \varrho_{k-1}} \to \mathcal{K}_{\pi_{k-1}+c_k, \varrho_{k-1} \setminus c_k}.$$

Now $\pi_{k-1} + c_k = \pi + \omega_{+,k} = \pi_k$ and it can be seen easily that $\varrho_{k-1} \setminus c_k = \varrho_k$.

Assume, now, that $\theta_k = -1$. In order that a_{c_k} is defined,

$$c_k \notin \pi_{k-1} + \varrho_{k-1} \subset \pi + \varrho + \omega_{k-1}.$$

But $c_k \notin \pi + \varrho$ by assumption and $c_k \notin \omega_{k-1}$, as W_k is admissible.

$$a_{c_k} : \mathcal{K}_{\pi_{k-1}, \varrho_{k-1}} \to \mathcal{K}_{\pi_{k-1}, \varrho_{k-1}+c_k}.$$

But $\omega_{+,k} = \omega_{+,k-1}$ and

$$\begin{aligned}
\pi_k &= \pi + \omega_{k-1} = \pi_{k-1}, \\
\varrho_k &= \varrho \setminus (\omega_{+,k} \setminus \omega_{-,k}) + (\omega_{-,k} \setminus \omega_{+,k}) \\
&= \varrho \setminus (\omega_{+,k-1} \setminus \omega_{-,k-1}) + (\omega_{-,k} \setminus \omega_{+,k}) + c_k \\
&= \varrho_{k-1} + c_k.
\end{aligned}$$

□

Lemma 3.3. *Assume $W = (W_2, W_1)$ to be admissible, denote by M_2 and M_1 the corresponding monomials and let $c \neq c'$ be indices. If $(W_2, a_c^\theta, a_{c'}^{\theta'}, W_1)$ is admissible, so is $(W_2, a_{c'}^{\theta'}, a_c^\theta, W_1)$ and*

$$\begin{aligned}
M_2 a_c^+ a_{c'}^+ M_1 &= M_2 a_{c'}^+ a_c^+ M_1 \\
M_2 a_c a_{c'} M_1 &= M_2 a_{c'}^+ a_c^+ M_1 \\
M_2 a_c a_{c'}^+ M_1 &= M_2 a_{c'}^+ a_c M_1 + \varepsilon_{x_c}(dx_{c'}) M_2 M_1.
\end{aligned}$$

For the proof combine lemmata 3.1 and 3.2.

4. Wick's Theorem

If S is a finite chain, denote by $\mathfrak{P}_2(S)$ the set of all pair partitions of S, i.e., the set of all

$$\mathfrak{p} = \left\{ (s_i, t_i), s_i > t_i; \{s_i, t_i\} \cap \{s_j, t_j\} = \emptyset \text{ for } i \neq j; \bigcup \{s_i, t_i\} = S \right\}.$$

If $\#S$ is odd, $\mathfrak{P}_2(S)$ is empty. Define

$$\langle a_c^\theta, a_{c'}^{\theta'} \rangle = \begin{cases} \varepsilon(c, c') & \text{for } \theta = -1,\ \theta' = +1, \\ 0 & \text{otherwise,} \end{cases}$$

writing for short

$$\varepsilon(c, c') = \varepsilon_{x_c}(dx_{c'}).$$

Definition 4.1. Assume

$$W = (a_{c_n}^{\theta_n}, \ldots, a_{c_1}^{\theta_1})$$

to be an admissible sequence, then define

$$\langle W \rangle = \sum_{\mathfrak{p} \in \mathfrak{P}_2([1,n])} \langle W \rangle_{\mathfrak{p}}.$$

If n is even and

$$\mathfrak{p} = \{(i(1), j(1)), \ldots, (i(n/2), j(n/2))\},$$

then

$$\langle W \rangle_{\mathfrak{p}} = \prod_{k=1}^{n/2} \left\langle a_{c_{i(k)}}^{\theta_{i(k)}}, a_{c_{j(k)}}^{\theta_{j(k)}} \right\rangle.$$

If n is odd, $\langle W \rangle = 0$.

As there is a 1–1 correspondence between $[1, n]$ and the chain $((c_1, \theta_1), \ldots, (c_n, \theta_n))$, we may write $\mathfrak{P}_2([1, n]) = \mathfrak{P}_2(W)$. We denote by $\mathfrak{P}_{20}(W)$ the set of those pair partitions where only pairs of the form

$$\langle a_c, a_{c'}^+ \rangle = \varepsilon(c, c')$$

occur. Then $\langle W \rangle_{\mathfrak{p}} = 0$ for $\mathfrak{p} \notin \mathfrak{P}_{20}(W)$ and

$$\langle W \rangle = \sum_{\mathfrak{p} \in \mathfrak{P}_{20}(W)} \langle W \rangle_{\mathfrak{p}}.$$

We want to investigate $\langle W \rangle_{\mathfrak{p}}$ for $\mathfrak{p} \in \mathfrak{P}_{20}(W)$. We define in $\mathfrak{p}$ a relation of nearest right neighborhood

$$(a_{b_1}, a_{b_2}^+) \rhd (a_{b_3}, a_{b_4}^+) \iff b_2 = b_3.$$

As for a given pair there is at most one nearest neighbor, the set $\mathfrak{p}$ splits into a a family of maximal chains of the form

$$\mathcal{C} = (a_1, a_2^+) \rhd (a_2, a_3^+) \rhd \cdots \rhd (a_{k-1}, a_k^+). \qquad (*)$$

To that chain corresponds the quantity

$$f_{\mathcal{C}} = \varepsilon(1,2)\,\varepsilon(2,3) \cdots \varepsilon(k-1,k)$$

$$f_{\mathcal{C}}(x_1, dx_2, \ldots, dx_k) = \varepsilon(x_1, dx_2)\varepsilon(x_2, dx_3) \cdots \varepsilon(x_{k-1}, dx_k)$$

$$\int_{2,\ldots,k} f_{\mathcal{C}}(x_1, dx_2, \ldots, dx_k)\,\varphi(x_2, \ldots, x_k) = \varphi(x_1, \ldots, x_k).$$

Proposition 4.1. *Any $\mathfrak{p} \in \mathfrak{P}_{20}(W)$ is the union of maximal chains $\mathcal{C}_l$ with respect to the relation $\rhd$. If*

$$\mathfrak{p} = \sum_l \mathcal{C}_l,$$

then

$$\langle W \rangle_{\mathfrak{p}} = \prod_l f_{\mathcal{C}_l}.$$

Example 4.1. If

$$W = (a_1, a_2^+, a_2, a_3^+, \dots, a_{k-1}, a_k^+),$$

then $\mathfrak{P}_{20}(W)$ consists of only one element, namely

$$\mathfrak{p} = \{(a_1, a_2^+), (a_2, a_3^+), \dots, (a_{k-1}, a_k^+)\}$$

and that is the only maximal chain, namely the chain $\mathcal{C}$ from equation $(*)$.

Example 4.2. If W is *antinormal* ordered

$$W = (a_1, \dots, a_k, a_{k+1}^+, \dots, a_{2k}^+),$$

then $\mathfrak{P}_{20}(W)$ consists of $k!$ elements, namely

$$\mathfrak{p}_\sigma = \{(a_1, a_{\sigma(1)}^+), \dots, (a_k, a_{\sigma(k)}^+)\},$$

where σ runs through all bijections $[1,k] \to [k+1, 2k]$. The maximal chains are all of length 1.

Definition 4.2. An admissible sequence is called *normal ordered* if it is of the form

$$(a_{c_n}^+, \dots, a_{c_{k+1}}^+, a_{c_k}, \dots, a_{c_1}).$$

Example 4.3. If W is normal ordered, then $\mathfrak{P}_{20}(W)$ is empty.

By interchanging the indices any admissible sequence can be normal ordered. As any interchanging of the indices in the intervals $[1,k]$ and $[k+1,n]$ does not change the monomial $a_{c_n}^+ \cdots a_{c_{k+1}}^+ a_{c_k} \cdots a_{c_1}$, we can define

$$:W: = a_{c_n}^+ \cdots a_{c_{k+1}}^+ a_{c_k} \cdots a_{c_1},$$

if $(a_{c_n}^+, \dots, a_{c_{k+1}}^+, a_{c_k}, \dots, a_{c_1})$ is any normal ordering of W. We may write

$$:W: = a_{\omega_+(W)}^+ a_{\omega_-(W)}$$

with $\omega_\pm(W)$ defined in definition 3.2.

Define by $\mathfrak{P}(S)$ the set of all partitions of a finite chain S into singletons and ordered pairs. If $\mathfrak{p} \in \mathfrak{P}(S)$, denote by $\mathfrak{p}' \subset \mathfrak{p}$ the subset of singletons ordered by the order of S and by $\mathfrak{p}'' \subset \mathfrak{p}$ the subset of ordered pairs.

Proposition 4.2. *If W is an admissible sequence*

$$W = (a_{c_n}^{\theta_n}, \dots, a_{c_1}^{\theta_1}),$$

then

$$M = a_{c_n}^{\theta_n} \cdots a_{c_1}^{\theta_1} = \sum_{\mathfrak{p} \in \mathfrak{P}([1,n])} \lfloor W \rfloor_{\mathfrak{p}}$$

with

$$\lfloor W \rfloor_{\mathfrak{p}} =: \prod_{i \in \mathfrak{p}'} a_{c_i}^{\theta_i} : \prod_{\{j,k\} \in \mathfrak{p}'', j>k} \langle a_{c_j}^{\theta_j}, a_{c_k}^{\theta_k} \rangle.$$

Proof. We proceed by induction. The case $n = 1$ is clear. We consider the mapping $\varphi : \mathfrak{P}([1,n]) \to \mathfrak{P}([1,n-1])$ defined by erasing n. If $\mathfrak{p} \in \mathfrak{P}([1,n])$ is of the form

$$\mathfrak{p} = \{n > r_k > \cdots > r_1\} + \sum_j \{s_j, t_j\}$$

with $s_j > t_j$, then

$$\varphi\mathfrak{p} = \{r_k > \cdots > r_1\} + \sum_j \{s_j, t_j\}.$$

If

$$\mathfrak{p} = \{r_k > \cdots > r_1\} + \{n, t\} + \sum_j \{s_j, t_j\}$$

then

$$\varphi\mathfrak{p} = \{r_k > \cdots > r_1\} + \{t\} + \sum_j \{s_j, t_j\}.$$

Assume $\mathfrak{q} \in \mathfrak{P}([1, n-1])$,

$$\mathfrak{q} = \{r_k > \cdots > r_1\} + \sum_j \{s_j, t_j\},$$

then

$$\varphi^{-1}\mathfrak{q} = \{\mathfrak{p}_0, \mathfrak{p}_1, \ldots, \mathfrak{p}_k\}$$

with

$$\mathfrak{p}_0 = \{n > r_k > \cdots > r_1\} + \sum_j \{s_j, t_j\}$$

$$\mathfrak{p}_i = (\{r_k > \cdots > r_1\} \setminus r_i) + \{n, r_i\} + \sum_j \{s_j, t_j\}$$

for $i = 1, \ldots, k$.

Assume

$$V = (a_{c_{n-1}}^{\theta_{n-1}}, \ldots, a_{c_1}^{\theta_1})$$

to be an admissible sequence and let L be the corresponding monomial. Then by induction hypothesis

$$L = \sum_{\mathfrak{q}\in\mathfrak{P}([1,n-1])} \lfloor V \rfloor_{\mathfrak{q}}.$$

Assume, now, that (a_c, V) is admissible. So $c \notin \{c_{n-1}, \ldots, c_1\}$. Denote again by $\mathfrak{q}'$ and $\mathfrak{q}''$ the subsets of singletons and pairs of $\mathfrak{q}$. Then

$$\begin{aligned} a_c \lfloor V \rfloor_{\mathfrak{q}} &= a_c : \prod_{i\in\mathfrak{q}'} a_{c_i}^{\theta_i} : \prod_{(j>k)\in\mathfrak{q}''} \langle a_{c_j}^{\theta_j}, a_{c_k}^{\theta_k} \rangle \\ &= \left(: \prod_{i\in\mathfrak{q}'} a_{c_i}^{\theta_i} : a_c + \sum_{i\in\mathfrak{q}'} \langle a_c, a_{c_i}^{\theta_i} \rangle : \prod_{l\in(\mathfrak{q}'\setminus i)} a_{c_l}^{\theta_l} : \right) \prod_{(j>k)\in\mathfrak{q}''} \langle a_{c_j}^{\theta_j}, a_{c_k}^{\theta_k} \rangle. \end{aligned}$$

If $\mathfrak{q}' = \{r_k > \cdots > r_1\}$, we obtain

$$a_c \lfloor V \rfloor_{\mathfrak{q}} = \lfloor W \rfloor_{\mathfrak{p}_0} + \cdots + \lfloor W \rfloor_{\mathfrak{p}_k} = \sum_{\mathfrak{p}\in\varphi^{-1}(\mathfrak{q})} \lfloor W \rfloor_{\mathfrak{p}}.$$

Sum over $\mathfrak{q}$ and obtain the formula for (a_c, V).

Assume (a_c^+, V) to be admissible, then $c \notin \omega_+(V)$. As

$$a_c^+ : \prod_{i\in\mathfrak{q}'} a_{c_i}^{\theta_i} : = : a_c^+ \prod_{i\in\mathfrak{q}'} a_{c_i}^{\theta_i} :,$$

we obtain that

$$a_c \lfloor V \rfloor_{\mathfrak{q}} = \lfloor W \rfloor_{\mathfrak{p}_0} = \sum_{\mathfrak{p}\in\varphi^{-1}(\mathfrak{q})} \lfloor W \rfloor_{\mathfrak{p}},$$

as for $i = 1, \ldots, k$ the quantity $\lfloor W \rfloor_{\mathfrak{p}_i}$ contains the factor $\langle a_c^+, a_{c_i}^{\theta_i} \rangle = 0$. □

Recall the definitions of the function Φ and the measure Ψ from equations (1) and (7). If M is a normal ordered monomial, then

$$\Psi M \Phi = \begin{cases} 1 & \text{for } M = 1, \\ 0 & \text{otherwise.} \end{cases}$$

One deduces from the last proposition

Theorem 4.1 (Wick's theorem). *If*

$$W = (a_{c_n}^{\theta_n}, \ldots, a_{c_1}^{\theta_1})$$

is an admissible sequence and M the corresponding monomial, then

$$\Psi M \Phi = \Psi a_{c_n}^{\theta_n} \cdots a_{c_1}^{\theta_1} \Phi = \langle W \rangle.$$

So $\langle W\rangle$ depends only of the monomial M, defined by W, and we may write $\langle W\rangle = \langle M\rangle$. From the last proposition we obtain furthermore

Theorem 4.2. *Assume*

$$W = (a_{cn}^{\theta_n}, \dots, a_{c1}^{\theta_1}),$$

to be an admissible sequence, and assume subsets I, J such that $I + J = [1, n]$, and let W_I and W_J be the restrictions of W to I resp. J. Denote by M, M_I, M_J the corresponding monomials. Then,

$$M = \sum_{I+J=[1,n]} :M_I: \langle M_J\rangle.$$

5. Duality

Fix a positive measure λ on X, denote by $e(\lambda)$ the corresponding measure on $\mathfrak{X}$ and write for short

$$e(\lambda)(dx_\alpha) = \lambda(x_\alpha) = \lambda_\alpha.$$

We define a scalar product on $\mathcal{K} = \mathcal{K}_s(\mathfrak{X})$ by

$$\begin{aligned}\langle f, g\rangle_\lambda &= \int \overline{f(x_\alpha)}\, g(x_\alpha)\, \lambda(dx_\alpha) \\ &= \overline{f(\emptyset)}\, g(\emptyset) + \sum_{n=1}^{\infty} \frac{1}{n!} \int \cdots \int \overline{f(x_1, \dots, x_n)}\, g(x_1, \dots, x_n)\, \lambda(dx_1, \dots, dx_n).\end{aligned}$$

Using equation 10, we obtain for $\varphi \in \mathcal{K}(X)$

$$\langle f, a^+(\varphi) g\rangle_\lambda = \langle a(\overline{\varphi}\lambda), g\rangle_\lambda,$$

where $\overline{\varphi}\lambda$ is the product of the function $\overline{\varphi}$ with the measure λ.

Extend the scalar product to measure valued functions.

Lemma 5.1. *One has*

$$\int_\alpha \overline{f(x_\alpha)}\, (a_c^+ g)(x_\alpha, dx_c)\, \lambda(dx_\alpha) = \int_\alpha \overline{(a_c f)(x_\alpha, x_c)}\, \lambda(dx_c)\, g(x_\alpha)\, \lambda(dx_\alpha)$$

or writing for short

$$\langle f, a_c^+ g\rangle_\lambda = \langle (a_c f)\lambda_c, g\rangle_\lambda.$$

Proof. Using the sum-integral-lemma

$$\int_\alpha \overline{f(x_\alpha)} \sum_{b\in\alpha} \varepsilon_{x_b}(dx_c)\, g(x_{\alpha\backslash c})\, \lambda(dx_\alpha)$$
$$= \int_\alpha \int_b \overline{f(x_{\alpha+b})}\, \varepsilon_{x_b}(dx_c)\, g(x_\alpha)\, \lambda(dx_{\alpha+b}).$$

Now

$$\int_b \overline{f(x_{\alpha+b})}\, \varepsilon_{x_b}(dx_c)\, \lambda(dx_b) = \overline{f(x_{\alpha+c})}\lambda(dx_c)$$

as

$$\int_c \int_b \overline{f(x_{\alpha+b})}\, \varepsilon_{x_b}(dx_c)\, \lambda(dx_b)\, \varphi(x_c) = \int_c \overline{f(x_{\alpha+c})}\, \lambda(dx_c)\, \varphi(x_c).$$

□

Definition 5.1. Assume

$$W = (a_{c_n}^{\theta_n}, \ldots, a_{c_1}^{\theta_1})$$

to be an admissible sequence, then define the formal adjoint sequence by

$$W^+ = (a_{c_1}^{-\theta_1}, \ldots, a_{c_n}^{-\theta_n}).$$

If M is the monomial corresponding to W, we denote by M^+ the monomial corresponding to W^+.

Recall definition 3.2 and call $\omega = \omega(W)$, etc.; then, by proposition 3.1,

$$M : \mathcal{K} \to \mathcal{K}_{\omega_+, \omega_- \backslash \omega_+}.$$

We may interpret M as a function $X_{\omega_- \backslash \omega_+} \to \mathcal{M}(X^{\omega_+})$ with values in the set of linear operators $\mathcal{K} \to \mathcal{K}$ and write

$$M = M\big((dx_i)_{i\in\omega_+}, (x_j)_{j\in\omega_-\backslash\omega_+}\big).$$

Then

$$\begin{aligned} M\lambda_{\omega_-\backslash\omega_+} &= (M\lambda_{\omega_-\backslash\omega_+})(dx_i)_{i\in\omega} \\ &= M\big((dx_i)_{i\in\omega_+}, (x_j)_{j\in\omega_-\backslash\omega_+}\big)\, \lambda\big((dx_k)_{k\in\omega_-\backslash\omega_+}\big) \end{aligned}$$

is a measure on X^ω with values in the linear operators $\mathcal{K} \to \mathcal{K}$. In a similar way,

$$M^+\lambda_{\omega_+\backslash\omega_-} = (M\lambda_{\omega_+\backslash\omega_-})(dx_i)_{i\in\omega}$$

is a measure on X^ω too.

By induction we prove, by the last lemma,

Theorem 5.1. *For $f, g \in \mathcal{K}$ we have*

$$\langle f, (M\lambda_{\omega_-\backslash\omega_+})g\rangle_\lambda = \langle (M^+\lambda_{\omega_+\backslash\omega_-})f, g\rangle_\lambda.$$

6. Meyer's formula

If W is normal ordered, then the corresponding monomial is of the form

$$M = a^+_{\sigma+\tau} a_{\tau+\upsilon},$$

for σ, τ, υ in finite index sets. A normal ordered sequence is always admissible. As it has been pointed out in the last section, $M\lambda_\upsilon$ is a measure on $X^{\sigma+\tau+\upsilon}$ with values in the linear operators $\mathcal{K} \to \mathcal{K}$. Assume a function

$$f = f(x_\sigma, x_\tau, x_\upsilon) \in \mathcal{K}(\mathfrak{X}^3)$$

symmetric in every variable $x_\sigma, x_\tau, x_\upsilon$ and consider

$$T(f) = \int_{\sigma,\tau,\upsilon} f(x_\sigma, x_\tau, x_\upsilon)\, a^+_{\sigma+\tau} a_{\tau+\upsilon}(dx_\sigma, dx_\tau, x_\upsilon)\, \lambda(dx_\upsilon)$$

or, written in an explicit way,

$$\begin{aligned} T(f) = \sum_{l,m,n=0}^{\infty} \frac{1}{l!m!n!} \int \cdots \int & f(x_1, \ldots, x_l; y_1, \ldots, y_m; z_1, \ldots, z_n) \\ & a^+(\varepsilon)(dx_1) \cdots a^+(\varepsilon)(dx_l) a^+(\varepsilon)(dy_1) \cdots a^+(\varepsilon)(dy_m) \\ & a(\varepsilon_{y_1}) \cdots a(\varepsilon_{y_m})\, a(\varepsilon_{y_1}) \cdots a(\varepsilon_{y_m})\, \lambda(dz_1) \cdots \lambda(dz_n). \end{aligned}$$

As f is of compact support, the sum contains only finitely many terms and the integral is well defined and yields an operator $\mathcal{K} \to \mathcal{K}$. That is the generalization of the Maassen-Meyer kernel representation generalized to locally compact spaces. We write for short

$$T(f) = \int_\upsilon f(\sigma, \tau, \upsilon)\, a^+_{\sigma+\tau} a_{t+\upsilon} \lambda_\upsilon.$$

We want to prove Meyers formula for the composition of two operators.

Theorem 6.1. *If f, g are two functions in $\mathcal{K}(\mathfrak{X}^3)$ symmetric in each variable, then*

$$T(f)\, T(g) = T(h)$$

with

$$h(\sigma,\tau,\upsilon) =$$

$$\sum \int_\kappa \lambda_\kappa f(\alpha_1,\alpha_2+\beta_1+\beta_2,\gamma_1+\gamma_2+\kappa)\, g(\kappa+\alpha_2+\alpha_3,\beta_2+\beta_3+\gamma_2,\gamma_3)$$

where the sum runs through all indices $\alpha_1,\ldots,\gamma_3$ *with*

$$\begin{aligned}\alpha_1+\alpha_2+\alpha_3 &= \sigma,\\ \beta_1+\beta_2+\beta_3 &= \tau,\\ \gamma_1+\gamma_2+\gamma_3 &= \upsilon.\end{aligned}$$

That is essentially Meyer's formula [9, p. 92]. The difference is mainly that his formula is formulated for sets of coordinates, whereas our formula deals with sets of indices of coordinates. So in our formula σ stands for $x_\sigma = (x_{s_1},\ldots,x_{s_l})$. Meyer indicates the formula for much more general functions. In order to generalize our formula to more complicated functions we had to use the extension theorems of measure theory. Meyer's statement is formulated for $X = \mathbb{R}$ and λ Lebesgue measure. It could be extended easily to locally compact X and a diffuse measure λ.

Proof. We define

$$\varepsilon(\alpha,\beta) = \langle a_\alpha, a_\beta^+\rangle = \sum_{\varphi\in B(\alpha,\beta)} \prod_{c\in\alpha} \varepsilon(c,\varphi(c)),$$

where $B(\alpha,\beta)$ is the set of all bijections $\varphi:\alpha\to\beta$. If $\#\alpha \neq \#\beta$, then $B(\alpha,\beta)=\emptyset$ and $\varepsilon(\alpha,\beta)=0$.

One shows easily that

$$\begin{aligned}\varepsilon(\alpha_1+\alpha_2,\beta) &= \sum_{\beta_1+\beta_2=\beta} \varepsilon(\alpha_1,\beta_1)\,\varepsilon(\alpha_2,\beta_2)\,,\\ \varepsilon(\alpha,\beta_1+\beta_2) &= \sum_{\alpha_1+\alpha_2=\alpha} \varepsilon(\alpha_1,\beta_1)\,\varepsilon(\alpha_2,\beta_2)\,.\end{aligned}$$

From there one concludes that

$$\varepsilon(\alpha_1+\alpha_2,\beta_1+\beta_2) = \sum \varepsilon(\alpha_{11},\beta_{11})\,\varepsilon(\alpha_{12},\beta_{21})\,\varepsilon(\alpha_{21},\beta_{12})\,\varepsilon(\alpha_{22},\beta_{22}), \quad (*)$$

where the sum runs through all indices $\alpha_{11},\ldots,\beta_{22}$ with

$$\begin{aligned}\alpha_{11}+\alpha_{12} &= \alpha_1, & \alpha_{21}+\alpha_{22} &= \alpha_2,\\ \beta_{11}+\beta_{12} &= \beta_1, & \beta_{21}+\beta_{22} &= \beta_2.\end{aligned}$$

We have

$$T(f)\,T(g) = \int \lambda_{\upsilon_1+\upsilon_2}\, f(\sigma_2,\tau_2,\tau_1)\, g(\sigma_1,\tau_1,\upsilon_1)\, a^+_{\sigma_2+\tau_2} a_{\tau_2+\upsilon_2} a^+_{\sigma_1+\tau_1} a_{\tau_1+\upsilon_1}\,,$$

where the integral runs over all (mutually disjoint) index sets $\sigma_1,\ldots,\upsilon_2$. Using theorem 4.2,

$$\begin{aligned} &a^+_{\sigma_2+\tau_2}\, a_{\tau_2+\upsilon_2}\, a^+_{\sigma_1+\tau_1}\, a_{\tau_1+\upsilon_1} \\ &\qquad\qquad = \sum a^+_{\sigma_2+\tau_2+\sigma_{11}+\tau_{11}}\, a_{\tau_{21}+\upsilon_{21}+\tau_1+\upsilon_1}\, \varepsilon(\tau_{22}+\upsilon_{22},\sigma_{12}+\tau_{12}) \end{aligned}$$

where the indices obey the conditions

$$\begin{aligned} \tau_{21}+\tau_{22} &= \tau_2, & \upsilon_{21}+\upsilon_{22} &= \upsilon_2, \\ \sigma_{11}+\sigma_{12} &= \sigma_1, & \tau_{11}+\tau_{12} &= \tau_1. \end{aligned}$$

Following $(*)$,

$$\begin{aligned} &\varepsilon(\tau_{22}+\upsilon_{22},\sigma_{12}+\tau_{12}) \\ &\qquad\qquad = \sum \varepsilon(\tau_{221},\sigma_{121})\,\varepsilon(\tau_{222},\tau_{121})\,\varepsilon(\upsilon_{221},\sigma_{122})\,\varepsilon(\upsilon_{222},\tau_{122}) \end{aligned}$$

with

$$\begin{aligned} \tau_{221}+\tau_{222} &= \tau_{22}, & \upsilon_{221}+\upsilon_{222} &= \upsilon_{22}, \\ \sigma_{121}+\sigma_{122} &= \sigma_{12}, & \tau_{121}+\tau_{122} &= \tau_{12}. \end{aligned}$$

Using the sum-integral lemma,

$$\begin{aligned} T(f)\,T(g) = \int &\lambda_{\upsilon_{21}+\upsilon_{221}+\upsilon_{222}+\upsilon_1} \\ &f(\sigma_2,\tau_{21}+\tau_{221}+\tau_{222},\upsilon_{21}+\upsilon_{221}+\upsilon_{222}) \\ &g(\sigma_{11}+\sigma_{121}+\sigma_{122},\tau_{11}+\tau_{121}+\tau_{122},\upsilon_1) \\ &\varepsilon(\tau_{221},\sigma_{121})\,\varepsilon(\tau_{222},\tau_{121})\,\varepsilon(\upsilon_{221},\sigma_{122})\,\varepsilon(\upsilon_{222},\tau_{122}) \\ &a^+_{\sigma_2+\tau_{21}+\tau_{221}+\tau_{222}+\sigma_{11}+\tau_{11}}\; a_{\tau_{21}+\upsilon_{21}+\tau_{11}+\tau_{121}+\tau_{122}+\upsilon_1}\,, \end{aligned}$$

where the integral runs over all indices. Put

$$\begin{aligned} \sigma_2 &= \alpha_1, & \sigma_{121} = \tau_{221} &= \alpha_2, & \sigma_{11} &= \alpha_3, \\ \tau_{21} &= \beta_1, & \tau_{222} = \tau_{121} &= \beta_2, & \tau_{11} &= \beta_3, \\ \upsilon_{21} &= \gamma_1, & \upsilon_{222} = \tau_{122} &= \gamma_2, & \upsilon_1 &= \gamma_3, \\ & & \sigma_{122} = \upsilon_{221} &= \kappa, & & \end{aligned}$$

where the equalities in the second column hold after integration. Define

$$\alpha_1 + \alpha_2 + \alpha_3 = \sigma, \quad \beta_1 + \beta_2 + \beta_3 = \tau, \quad \gamma_1 + \gamma_2 + \gamma_3 = \upsilon,$$

and obtain the theorem using the sum-integral lemma again. □

Acknowledgment

The author wants to thank Uwe Franz for carefully reading the manuscript and indicating some misprints and errors.

References

1. L. Accardi, Y.-G. Lu, I. V. Volovich, "White noise approach to classical and quantum stochastic calculus", Preprint 375, Centro Vito Volterra, Universita Roma 2 (1999).
2. L. Accardi, Y.-G. Lu, I. V. Volovich, "Quantum theory and its stochastic limit", Springer 2002.
3. S. Attal, "Problemes d'unicite dans les representations d'operateurs sur l'espace de Fock", *Seminaire de Probabilites XXVI*, *LNM* 1526, 1992, pp. 617–632.
4. N. Bourbaki, *Integration*, Paris, 1965, chap. 1–4.
5. N. Bourbaki, *Integration*, Paris, 1965, chap. 5.
6. J. M. Lindsay, "Quantum and non-causal stochastic calculus", *Prob. Theory Relat. Fields* **97**, pp. 65–80 (1993).
7. J. M. Lindsay, H. Maassen, *An integral kernel approach to noise*, *LNM* 1303, 1988, pp. 192–208.
8. H. Maassen, "Quantum Markov processes on Fock space described by integral kernels", in *LNM* 1136, 1985, pp. 361–374.
9. P. A. Meyer, *Quantum Probability for Probabilists*, *Lecture Notes in Mathematics* 1538, Springer, Berlin, Heidelberg, 1993.
10. N. Obata, *White noise calculus and Fock space*, *LNM* 1577, Springer, 1994.
11. K. R. Parthasarathy, *An Introduction to Quantum Stochastic Calculus*, Birkhaeuser, Basel, Boston, Berlin, 1992.
12. L. Schwartz, *Theorie des distributions I*, Herrmann, Paris, 1951.
13. W. von Waldenfels, "Continous Maassen kernels and the inverse oscillator", *Seminaire des Probabilites XXX*, *LNM* 1626, Springer, 1996.
14. W. von Waldenfels, "Continuous kernel processes in quantum probability", *Quantum Probability Communications* vol. XII, World Scientific, 2003, pp. 237–260.
15. W. von Waldenfels, "Symmetric differentiatiation and Hamiltonian of a quantum stochastic process", *Infinite dimensional analysis, quantum probability and related topics* **8**, 73–116 (2005).

WILHELM VON WALDENFELS
Institut für Angewandte Mathematik, Universität Heidelberg,
Im Neuenheimer Feld 294, 69120 Heidelberg, Germany
Wilhelm.Waldenfels@T-Online.de

FROM THE GEOMETRY OF PARABOLIC PDE TO THE GEOMETRY OF SDE*

JEAN-CLAUDE ZAMBRINI (*GFM-UL, Lisboa*)

We consider various notions of integrability in classical, quantum and stochastic dynamics. In the first field there is a plethora of definitions, sometimes not equivalent. In the two last ones, there are no universally accepted definitions, although a number of recent works (in quantum chaos, for example) suggest that it would be useful to have one.

We show here that a deformation of the classical notion of integrability underlying contact geometry provides a new perspective on stochastic integrability, with potential consequences in quantum mechanics.

1. Introduction

This is an essay on the notion of integrability in classical, quantum and stochastic dynamics. As a matter of fact, the status of this notion is quite distinct in these 3 contexts. We shall start with a short survey of the simplest version of integrability, the one due to J. Liouville in classical dynamics. The classical notion required for our "transversal" purpose is, in fact, more general than Liouville's one, inspiring Symplectic Geometry, or even more general than the one involved in Poisson geometry. It is the contact geometrical one, expressed in terms of Cartan's ideal of differential forms.

In quantum mechanics, there are still discussions about what should be an integrable system. We shall explain why.

Regarding the notion of dynamics provided by Stochastic Analysis, starting from Itô's theory of Stochastic Differential Equations (SDE), we will mention some of the difficulties one meets with when trying to define integrability. It seems, in particular, that despite the remarkable progress of Stochastic Analysis during the last 25 years (cf. [1] for example) a counterpart of the classical Frobenius Theorem for stochastic differential equations is not yet available. However, there is a dual version of Frobenius for differentiable forms and we shall see that a stochastic counterpart of this one could be within reach, at least in some special, dynamical contexts.

*This work is supported by the project POCI/MAT/55977/2004.

Our vital lead, in this expository review, will be the idea that an appropriate deformation of classical integrability notions should allow to shed a new light on their partly missing quantum and stochastic counterparts.

We shall conclude with a list of examples, open problems and prospects of the stochastic deformation strategy summarized here.

2. On classical integrability

Let us consider the following Hamiltonian system on the open set $M \subset \mathbb{R}^{2n}$ where n is called the number of degrees of freedom:

$$\dot{q}^i = \frac{\partial h}{\partial p_i}, \; \dot{p}_i = -\frac{\partial h}{\partial q^i}. \tag{1}$$

The dot denotes the derivative with respect to time, and $h = h(q,p,t)$ is the (scalar) Hamiltonian observable. Let us denote by

$$\{f,g\} = \frac{\partial f}{\partial q^i}\frac{\partial g}{\partial p_i} - \frac{\partial g}{\partial q^i}\frac{\partial f}{\partial p_i} \tag{2}$$

the Poisson bracket of the observable f and g (where we have used Einstein summation convention). A first integral of the systems is a function $n = n(q,p,t)$ constant along the solutions of equation (1), i.e., satisfying

$$\frac{\partial n}{\partial t} + \{h,n\} = 0. \tag{3}$$

In particular, for conservative systems, namely those with a time independent Hamiltonian $h = h(q,p)$, this observable is itself a first integral.

Liouville's Theorem [2] states that if the above mentioned Hamiltonian system admits n first integrals $n_i(q,p)$, $1 \leqslant i \leqslant n$, in involution (i.e., $\{n_i, n_j\} = 0$, $\forall i,j$) and if the functions n_i are functionally independent on compact level sets, then there is a canonical coordinates transformation (to "action-angle" variables) on a n-dimensional torus where the flow becomes linear in time.

This "complete" integrability has been generalized in many ways. The first integrals can be time-dependent, the Lie algebra of the first integrals can be noncommutative (then the system is called "super integrable"), etc. ... Cf. [2].

Liouville's Theorem is at the origin of the principle that completely integrable systems can be solved "by quadrature" in terms of their first integrals.

The non-degeneracy of the bracket (2) allows to endow M with a non-degenerate 2-form named after Liouville:

$$\Omega = dp_i \wedge dq_i \tag{4}$$

which is closed by the Jacobi identity: $d\Omega = 0$. It gives a symplectic structure on M and defines the Poincaré 1-form

$$\omega = p_i\, dq^i \text{ such that } \Omega = d\omega. \tag{5}$$

Using it, we have

$$\Omega(X_g, X_f) = \{f, g\} = X_g(f),$$

defining the Hamiltonian vector fields

$$X_g = \frac{\partial g}{\partial p_i}\frac{\partial}{\partial q^i} - \frac{\partial g}{\partial q^i}\frac{\partial}{\partial p_i}. \tag{6}$$

It is often useful to extend the dimension of M by two, by introducing the extra pair of canonical variables time and (minus) energy: $(t, -E)$. Then the starting 1-form becomes the one of Poincaré-Cartan:

$$\omega_{pc} = p_i dq^i - E\, dt \tag{5'}$$

with the associated extended versions of Hamiltonian vector field and Poisson bracket.

Conservative Hamiltonian systems with one degree of freedom (i.e., $n = 2$) are always integrable by quadrature. Their trajectories lie entirely on the level sets of the Hamiltonian and provide the simplest example of the reduction procedure. More generally, when the Hamiltonian system has l independent first integrals in involution it can be reduced to a system with $(n - l)$ degrees of freedom.

It is known at least since the seventies that generic Hamiltonians systems are not integrable. There is an obvious relation (but far from trivial to formulate exactly) between integrability and regularity of the trajectories of Hamiltonian flows. No one would expect a (deterministic) chaotic dynamics from an integrable Hamiltonian system, for example.

The approach of integrability needed for our purpose is not as familiar as Liouville's one. It is inspired by the geometric theory of partial differential equation, for us the Hamilton-Jacobi equation associated with the Hamiltonian system (1):

$$-\frac{\partial S}{\partial t} + h(q, -\nabla S, t). \tag{7}$$

Let us recall that to each Hamiltonian system are, actually, associated two Hamilton-Jacobi equations, adjoint to each other with respect to the time parameter (cf. [3]). Our choice of (7) is not the traditional one but is natural for the stochastic deformation described afterwards (cf. [4] for the justification).

We are going to specialize to one degree of freedom (and even, later on, to an elementary class of Hamiltonians h) without lack of generality for the method advocated.

So let us consider $\{q,t\}$, the coordinates of the base configuration manifold of independent variables for equation (7), and $\{S\}$ the one of the dependent variable. The coordinates in the jet bundle of order 1, denoted J^1, are defined by

$$\left\{q, t, S, \frac{\partial S}{\partial q}, \frac{\partial S}{\partial t}\right\} \tag{8}$$

but regarded as independent variables. It is, therefore, much safer to choose a new label for the 2 last variables,

$$p = -\frac{\partial S}{\partial q}, \; E = \frac{\partial S}{\partial t}. \tag{9}$$

In the 5 dimensional space J^1, our PDE (7) becomes a 4 dimensional manifold, specified by a function

$$f(q,t,S,p,E) \equiv h(q,p,t) - E = 0 \tag{10}$$

called the characteristic of Hamilton-Jacobi equation.

By definition, for a given S, the section of S lifts up the base configuration manifold into the jet space, namely

$$\left\{q, t, S(q,t), \frac{\partial S}{\partial q}(q,t), \frac{\partial S}{\partial t}(q,t)\right\}. \tag{11}$$

Notice that $S = S(q,t)$ solves Hamilton-Jacobi equation (7) iff the 2-dimensional "section" (11) lies entirely in the 4-dimensional manifold (10) specifying this equation.

A symmetry of Hamilton-Jacobi equation will be a mapping of sections into other sections but staying in its defining manifold (10).

Lie characteristic equations for any PDE defined by f as in (10) are given by

$$\begin{aligned} q' = \frac{\partial f}{\partial p}, \; t' = -\frac{\partial f}{\partial E}, \; p' = -\frac{\partial f}{\partial q} - p\frac{\partial f}{\partial S}, \; E' = \frac{\partial f}{\partial t} - E\frac{\partial f}{\partial S}, \\ S' = f - p\frac{\partial f}{\partial p} - E\frac{\partial f}{\partial E}, \end{aligned} \tag{12}$$

where the $'$ denotes the derivative with respect to a parameter, say u. Specializing equation (10) to the elementary class of Hamiltonians $h(q,p,t) = \frac{1}{2}p^2 + V(q,t)$, we obtain

$$\begin{aligned} \dot{q} &= p,\ \dot{p} = -\frac{\partial V}{\partial q},\ \dot{h} = \frac{\partial V}{\partial t}, \\ \dot{S} &= -L(\dot{q},q,t) = -\Big(\frac{1}{2}\dot{q}^2 - V(q,t)\Big), \end{aligned} \tag{13}$$

where $\cdot$ denotes now $\frac{d}{dt}$, since the parameter u can be identified with t. The last equation provides us with the definition of the Lagrangian L (the minus sign is due to our unusual choice of Hamilton-Jacobi).

The vector field on J^1 corresponding to the characteristic (12), denoted by X_f, is called a contact vector field ([5]):

$$\begin{aligned} X_f = {} & \frac{\partial f}{\partial p}\frac{\partial}{\partial q} - \frac{\partial f}{\partial E}\frac{\partial}{\partial t} + \left(f - p\frac{\partial f}{\partial p} - E\frac{\partial f}{\partial E}\right)\frac{\partial}{\partial S} \\ & - \left(\frac{\partial f}{\partial q} + p\frac{\partial f}{\partial S}\right)\frac{\partial}{\partial p} + \left(\frac{\partial f}{\partial t} - E\frac{\partial f}{\partial S}\right)\frac{\partial}{\partial t}. \end{aligned} \tag{14}$$

Like an Hamiltonian vector field, X_f is determined uniquely by the (contact) Hamiltonian f. But contact geometry is distinct from the Symplectic one mentioned before. Instead of Liouville 2-form Ω, its fundamental tool is the following contact 1-form, which takes into account the "extra" dimension S of J^1:

$$\omega = \omega_{pc} + dS, \tag{15}$$

any contact Hamiltonian being defined from it by

$$\omega(X_f) = f. \tag{16}$$

A bracket $\{\cdot,\cdot\}_L$ called after Lagrange (some authors prefer "Jacobi bracket") transfers the Lie algebra structure of contact vector fields to the space $C^\infty(J^1)$ of contact Hamiltonians:

$$\omega([X_g, X_f]) = \{f,g\}_L. \tag{17}$$

Only when f and g are independent of the variable S, the two contact vector fields reduce to (extended) Hamiltonian vector fields and $\{\cdot,\cdot\}_L$ to the associated Poisson bracket.

In the contact geometric context, for a given f defining the PDE (7), like in (10), an infinitesimal symmetry n of Hamilton-Jacobi will be defined as a contact Hamiltonian n such that

$$\{f, n\}_L = 0. \tag{18}$$

For our elementary class of Hamiltonians involved in (13) it is a simple exercise to verify that

$$n(q, t, S, p, E) = X(q, t)p - T(t)E - \phi(q, t) \tag{19}$$

is such an infinitesimal symmetry of the associated Hamilton-Jacobi equation when its coefficients X, T and ϕ satisfy the following relations, known as determining equations in Lie group theory:

$$\dot{T} = 2\frac{\partial X}{\partial q}, \quad \frac{\partial X}{\partial t} = \frac{\partial \phi}{\partial q}, \quad \frac{\partial \phi}{\partial t} = -\dot{T}V - X\frac{\partial V}{\partial q} - T\frac{\partial V}{\partial t}. \tag{20}$$

The most appropriate formulation of these ideas about symmetry and integrability of a PDE, in our stochastic deformation perspective, is E. Cartan's one in terms of ideal of differential forms [6].

For Hamilton-Jacobi with the above elementary class of Hamiltonians, one writes

$$\begin{aligned} \omega &= p\,dq - E\,dt + dS \\ \beta &= \left(E - \frac{1}{2}p^2 - V(q,t)\right) dq\,dt. \end{aligned} \tag{21}$$

The first form is the contact 1-form (15), involving the extended Poincaré-Cartan form (5′). The 2-form β (where the wedge product is understood) defines our Hamilton-Jacobi equation.

The set Λ of all differential forms over any manifold form an algebra, whose multiplication is the wedge (or hook) product. An ideal I of the algebra is such that the result of the multiplication of elements of I by any element of the algebra is in I.

When a PDE (or system of PDEs) is expressed as an ideal I, Cartan has shown that I must be closed with respect to the exterior derivative d. Then I is called a differential ideal. The reason is that we must be sure that the integrability conditions associated with the PDE are, indeed, satisfied. Our ideal (21) is not closed, but the remedy for that is simple: just add $d\omega$. Then

$$I = \begin{cases} \omega = p\,dq - E\,dt + dS \\ \Omega = d\omega = dp\,dq - dE\,dt \\ \beta = \left(E - \frac{1}{2}p^2 - V(q,t)\right) dq\,dt \end{cases} \tag{22}$$

is a differential ideal. Its "sectioned" forms, traditionally denoted by a tilde, are annuled by the base configuration submanifold of coordinates (q, t). For example, $\widetilde{\Omega} = d\widetilde{\omega} = 0$ implies the existence of $S = S(q, t)$ s.t. the relations (9) hold true.

In our five variables space J^1 with coordinate (q, t, S, p, E) the generator of a symmetry should be, a priori, defined as a vector field

$$X_n = X_n^q \frac{\partial}{\partial q} + X_n^t \frac{\partial}{\partial t} + X_n^S \frac{\partial}{\partial S} + X_n^p \frac{\partial}{\partial p} + X_n^E \frac{\partial}{\partial E} \tag{23}$$

with coefficients to be determined such that

$$\mathcal{L}_{X_n}(I) \subseteq I \tag{24}$$

where $\mathcal{L}_{X_n}$ denotes the Lie derivative along X_n. In fact, the appropriate coefficients of X_n are precisely given by the characteristic associated with the contact Hamiltonian f in (14), for $f = n$ as in (19), justifying our present notations.

3. Quantum integrability

Even when we restrict ourselves to the quantization of the elementary class of Hamiltonians of §2,

$$h(q, p, t) = \frac{1}{2}p^2 + V(q, t), \tag{25}$$

one faces a puzzling situation when trying to define the notion of quantum integrability.

The first idea is to try to define a quantum version of Liouville Theorem:

For a given Hamiltonian observable H densely defined on $\mathcal{H} = L^2(\mathbb{R}^n)$, we could say that this system is integrable if there are n "independent" first integral observables N_i, $1 \leqslant i \leqslant n$, i.e., such that $[H, N_i] = 0$ and $[N_i, N_j] = 0$, $1 \leqslant i, j \leqslant n$.

But J. Von Neumannn [7] has proved that the Hamiltonian H and the other first integrals N_i can be all expressed as functions of a single observable θ of the system so that any naive version of functional independence of those first integrals is ruined.

Interesting definitions of quantum integrability have been, nevertheless, put forward. A good description of the present state of knowledge on this issue is given in [8], together with a sufficient integrability criterion (lack of continuous spectrum for H). One feels, however, that many other quantum systems should also be integrable, like the free particle for example, i.e., $V = 0$ in (25).

The present consensus is that, in strong contrast with classical systems, "most" quantum systems should be integrable (because of the central role of the Spectral Theorem) and that no future definition of quantum integrability could imply the integrability of the corresponding classical systems.

This circle of ideas gained a new relevance with the hope that a quantum Hamiltonian whose spectrum is made of the zeros of the Riemann function should have a non integrable classical counterpart. And more generally with the study of the link between integrability of quantum Hamiltonian systems and the distribution of its energies.

For more about this issue, cf. [9].

4. On stochastic integrability

Consider the following system of stochastic differential equations (SDE) for a diffusion process $z_t = (z_t^i)$:

$$dz_t = \widetilde{B}(z_t, t)dt + \widetilde{G}(z_t, t)dw_t, \quad t \in I, \tag{26}$$

where $w_t = (w_t^i)$ are i-copies of one dimensional Wiener process, and $\widetilde{B} : \mathbb{R}^n \times I \to \mathbb{R}^n$, $\widetilde{G} : \mathbb{R}^n \times I \to \mathbb{R}^n \times I$ are given. Such SDEs have been introduced by K. Itô, and regarded as stochastic perturbation of systems of first order differential equations like our Hamiltonian system (1).

Conditions on $\widetilde{B}$ and $\widetilde{G}$ needed to assure existence and uniqueness of the solution of equation (26) have been known for a long time (cf. [10], for example).

The modern theory founded on SDE, Stochastic Analysis, has been developed (impressively) as a first order theory. Although for ODE, the equivalence between the system (1) and, say, the second order equation of Euler-Lagrange needed for Lagrangian mechanics is rather obvious, this is far from being the case of equation (26). But, classically, the dynamical content of the theory always appears in second order equations. This may explain, in part, why it is often so hard to relate results of Stochastic Analysis with problems inspired by theoretical and mathematical physics.

Interestingly it seems that there is no general notion of integrability for systems of SDEs like (26). Of course, when $\widetilde{B}$ (the "drift") and $\widetilde{G}$ are linear, equation (26) is explicitly integrable. For $\widetilde{G}$ independent of $q = z_t$, our main interest here, the solution can, in fact, be explicitly expressed in terms of the solution of the corresponding deterministic linear equation ($\widetilde{G} = 0$ in

(26)). This (rare) circumstance should certainly be a sufficient condition for a definition of integrability of SDE. Applying any diffeomorphism $\theta : \mathbb{R}^n \to \mathbb{R}^n$ to such a solution together with Itô's formula we could produce many "integrable" SDEs. Still no general (a priori) criterion seems to be known.

Our thesis is that the difficulties we face in looking for such a definition of stochastic integrability are very similar to the ones experienced in trying to define quantum integrability. To illustrate this idea, we suggest to:

(1) Restrict the class (26) of SDEs to those qualitatively closer to Hamiltonian equations (1), in order to preserve as much as possible their dynamical content.
(2) Use the probability measures associated with all the solutions of this class of SDE to deform the differential ideal (22) containing the symmetries and integrability conditions of the classical Hamiltonian-Jacobi equation (7), for the elementary class of Hamiltonians of §2.
(3) Choose, for (2), a quantum-like probabilistic deformation so that, after coming back in Hilbert space $\mathcal{H}$, we could infer a reasonable definition of quantum integrability.

The rest of this paragraph will describe some recent progress along this line of stochastic deformation, resulting from joint works with P. Lescot [11].

Our starting deformation of Hamilton-Jacobi equation (7) for the elementary class of Hamiltonians involved in (13) will be:

$$-\frac{\partial S}{\partial t} + \frac{1}{2}\left(\frac{\partial S}{\partial q}\right)^2 - V(q,t) - \frac{\hbar}{2}\frac{\partial^2 S}{\partial q^2} = 0. \tag{27}$$

This (uniformly) parabolic equation goes under the name of Hamilton-Jacobi-Bellman equation, our deformation parameter being denoted by $\hbar > 0$.

In analogy with (9), we define two variables by

$$B = -\frac{\partial S}{\partial q}, \quad E = -\frac{\partial S}{\partial t}. \tag{28}$$

B will become a drift later on. In equation (26) the drift is, of course, a given function of q and t, and this is why we have added a $\tilde{}$ at the top of B. But in (28), B and E are two independent variables, like in J^1. We shall keep such convention until the end of this exposition.

The deformation of the ideal (22), relevant to Hamilton-Jacobi-Bellman equation is given by

$$I_{HJB} = \begin{cases} \omega = B\,dq + E\,dt + dS \equiv \omega_{pc} + dS \\ \Omega = d\omega = dB\,dq + dE\,dt \\ \beta = \left(E + \frac{1}{2}B^2 - V\right) dq\,dt + \frac{\hbar}{2} dB\,dt. \end{cases} \tag{29}$$

There are a few changes of signs in (29) with respect to the classical ideal (22). They are expected because from (22) to (29) the real time parameter of Hamiltonian mechanics became the "imaginary time" one needs to make probabilistic sense of our $\hbar$-deformation.

Like their classical counterpart, the sectioned forms of (29) are annuled on the 2-submanifold of coordinates (q,t).

Calling N a symmetry generator defined exactly like (23), with unknown coefficient denoted by N^q, N^t, N^S, N^B, N^E, such that

$$\mathcal{L}_N(I_{HJB}) \subseteq I_{HJB}, \tag{30}$$

we characterize in this way a Lie algebra $\mathcal{G}$. It contains an infinite dimensional subalgebra originated in the observation that the change of variable

$$S(q,t) = -\hbar \ln \eta(q,t) \tag{31}$$

in equation (27) reduces this one to the linear heat equation

$$\hbar \frac{\partial \eta}{\partial t} = -\frac{\hbar^2}{2} \frac{\partial^2 \eta}{\partial q^2} + V\eta \equiv H\eta \tag{32}$$

and so the superposition principle applies to it. The most interesting part of $\mathcal{G}$ is the supplement $\mathcal{N}$ of the above subalgebra. When $V = 0$ in equation (27), for example, the dimension of $\mathcal{N}$ is 6, each element of its basis being associated with a symmetry of the system described by I_{HJB} (cf. [12]). Still, it is the symmetry underlying the change of variable (31) which is at the origin of our stochastic deformation method, since probabilistic interpretations of equation (32) have been known for a long time.

Given our "deformed" differential ideal (29), the first step is purely algebrical, however, and does not appeal for any probability whatsoever:

Theorem 4.1. *Let $N \in \mathcal{N}$ be a symmetry generator, with coefficients denoted as before, for the ideal I_{HJB} of (29). We shall call Lagrangian L the following function of 3 (independent) variables:*

$$(q, B, t) \mapsto L(q, B, t) = \frac{1}{2}B^2 + V(q,t). \tag{33}$$

Then the following Lie dragging relations hold:

$$(1)\ \mathcal{L}_N(\omega_{pc}) = -dN^s, \tag{34}$$

$$(2)\ \mathcal{L}_N(\Omega) = 0, \tag{35}$$

$$(3)\ \mathcal{L}_N(L) + L\frac{dN^t}{dt} = -DN^s, \tag{36}$$

where D is the differential operator defined by

$$D = \frac{\partial}{\partial t} + B\frac{\partial}{\partial q} + \frac{\hbar}{2}\frac{\partial^2}{\partial q^2}. \tag{37}$$

The proof of this Theorem is not hard, but quite long. It is given in [11]. We shall concentrate here on its interpretation in our perspective: the integrability of Hamilton-Jacobi-Bellman equation (27).

An interesting aspect of the proof of Theorem 4.1 is that it requires the introduction of the following counterpart of the symmetry contact Hamiltonian (19), associated here to $N \in \mathcal{N}$:

$$F_N(q,t,S,B,E) = X_N(q,t)\,B + T_N(t)E - \phi_N(q,t) \tag{38}$$

where $X_N = N^q$, $T_N = N^t$, $\phi_N = -N^s$ in terms of the above mentioned coefficients of our symmetry generator N. The corresponding deformation of the determining equation (20) is as follows:

$$\dot{T}_N = 2\frac{\partial X_N}{\partial q},\ \frac{\partial X_N}{\partial t} = \frac{\partial \phi_N}{\partial q},\ \frac{\partial \phi_N}{\partial t} + \frac{\hbar}{2}\frac{\partial^2 \phi_N}{\partial q^2} = \dot{T}_N V + X_N\frac{\partial V}{\partial q} + T_N\frac{\partial V}{\partial t}. \tag{39}$$

Although Theorem 4.1 does not have, a priori, a probabilistic meaning, the presence of the operator (37) is showing us the way to remedy this:
If B could be interpreted as the drift of a diffusion process like z_t in (26) (with $n = 1$, $\widetilde{G} = \hbar$) then the differential operator (37) would coincide with

$$\frac{\partial}{\partial t} + \mathcal{A}, \tag{40}$$

where $\mathcal{A}$ is called infinitesimal generator of the process. Since (26) is a (SD) Equation, the drift should, in this case, not be a variable independent of q and t, like in Theorem 4.1, but a given function of those two variables. Now this is precisely what the section $\sim$ of the S variable, both for the classical Hamilton-Jacobi equation (7) (cf. (11)) and its deformation (27), was all about. Moreover, using (31) and (28), provided by the geometry of I_{HJB}, we know that the dependence should be of the form:

$$\widetilde{B}(q,t) = \hbar \frac{\partial}{\partial q} \ln \eta(q,t), \quad \widetilde{E}(q,t) = \hbar \frac{\partial}{\partial t} \ln \eta(q,t). \tag{41}$$

To promote our variable q at time t to a diffusion process z_t with drift and energy given by (41), we only need to assume that the underlying solution η of (32) is positive. From now on, we shall denote by $\widetilde{D}$ the stochastic derivation operator (37) involving the above drift $\widetilde{B}$, namely the infinitesimal generator of z_t, built in term of such a positive η.

To the geometrical operation of section, needed here to construct z_t, corresponds an homomorphism of Lie algebra. Indeed, pick a symmetry generator as in Theorem 4.1. Then form the semigroup, for $\beta \in \mathbb{R}$,

$$e^{-\beta N} : (q,t,S,B,E) \to (q_\beta, t_\beta, S_\beta, B_\beta, E_\beta).$$

Defining $\eta_\beta(q_\beta, t_\beta)$ by $e^{-\frac{1}{\hbar}S_\beta}$, in conformity with (31), we observe that η_β solves (32) as well, in the variables (q_β, t_β). Then one can define a "reduced" symmetry generator $\widetilde{N}$ so that

$$e^{\beta \widetilde{N}} : \eta \mapsto \eta_\beta$$

and check that $N \to \widetilde{N} = N^q \frac{\partial}{\partial q} + N^t \frac{\partial}{\partial t} - \frac{1}{\hbar} N^S$ is a group homomorphism. This means that as long as our probabilistic interpretation of the ideal of HJB equation is concerned, nothing will be lost by such a reduction to the (q,t) submanifold.

This submanifold is also the one on which we can randomize (Lie) characteristic equations (13) along the irregular paths $t \mapsto z_t = q$, for our elementary class of Hamiltonians:

Theorem 4.2. *Let η a positive solution of (32), for $t \in I$ an interval of $\mathbb{R}$, and z_t, $t \in I$, a solution of the SDE (26) for the drift $\widetilde{B}$ of (41) and $\widetilde{G} = \hbar^{\frac{1}{2}}$. Then the following stochastic characteristic equations hold almost everywhere:*

$$\begin{aligned} &\widetilde{D} z_t = \widetilde{B}, \ \widetilde{D}\widetilde{B} = \frac{\partial V}{\partial q}(z_t,t), \ \widetilde{D}\widetilde{E} = \frac{\partial V}{\partial t}(z_t,t), \\ &\widetilde{D}\widetilde{S} = -L(z_t, \widetilde{B}, t). \end{aligned} \tag{42}$$

We shall consider mainly the last equation of (42), crucial for our purpose. After time integration and using a relation often called Dynkin formula, we obtain the action

$$\widetilde{S}(z_t,t) = E_t \int_t^u \left(\frac{1}{2}\widetilde{B}^2(z_\tau,\tau) + V(z_\tau,\tau) \right) d\tau \, + \, E_t \, \widetilde{S}(z_u,u), \qquad (43)$$

where E_t denotes the conditional expectation given z_t, and the integrand coincides with our Lagrangian (33).

The results of Theorem 4.1, after sectioning to the (q,t) submanifold describe, in particular, all the symmetries of the action (43). Before elaborating on this aspect, let us observe that this action, which is a function of z_t, can also be regarded as a (Boltz) functional J whose domain D_J contains all diffusions $X_\bullet$ absolutely continuous w.r.t. Wiener's measure and solving a SDE of the form (26), with fixed $\widetilde{G} = \hbar^{\frac{1}{2}}$ but arbitrary drift $\widetilde{B}$. Defining $X \in \mathcal{D}_J$ as an extremal of J if $E[\nabla J[X](\delta X)] = 0 \, \forall \, \delta X : \Omega_0 \to \chi$, where $\Omega_0 = \{\omega \in C(\mathbb{R}^+, \mathbb{R}) \text{ s.t. } \omega(0) = 0\}$ is the Wiener space, χ its Cameron-Martin subspace of absolutely continuous paths with square integrable derivative and $\nabla J[X](\delta X)$ denotes the directional derivative of J at X in the direction δX, one shows that the diffusion z_t of Theorem 4.2 is indeed extremal of J and satisfies (cf. [13]) a.s.:

$$\begin{cases} \widetilde{D}\widetilde{D}z_\tau = \nabla V(z_\tau), \quad t \le \tau \le u, \\ z_t = q, \quad \widetilde{D}z_u = -\nabla \widetilde{S}(z_u,u). \end{cases} \qquad (44)$$

Now, by construction, $\widetilde{S}$ solves the HJB equation (27). Taking $\frac{\partial}{\partial q}$ of this equation and introducing the definitions (28) of $\widetilde{B}$ one checks that the resulting relation coincides with (44), the boundary conditions being also satisfied. The other relations of (42) follow as easily.

So the diffusion z_t of Theorem 27 (or, equivalently, the HJB equation associated with it) satisfies the stochastic least action principle (cf. [4] for more on this) underlying the presence of the variable S in the ideal I_{HJB}.

For the geometry of the action (43) we rely, in particular, on the last claim of Theorem 4.1.

Let us consider (36) when $\hbar = 0$. By construction, this should essentially correspond to a geometrical statement of classical dynamics. Indeed it does: This is exactly the invariance property of the fundamental (action) integral under its symmetry group of transformations on the (q,t) manifold, which is the hypothesis of Noether's famous Theorem (cf. [14] p.204). The r.h.s. of (36) is, in general, not zero: one says that the action is invariant modulo the divergence term $\frac{d}{dt}N^S$.

Classically, i.e., for $\hbar = 0$, the invariance condition (36) means explicitly,

$$N^q \frac{\partial L}{\partial q} + N^t \frac{\partial L}{\partial t} + \left(\frac{d}{dt}N^q - \dot{q}\dot{N}^t\right)\frac{\partial L}{\partial \dot{q}} + L\dot{N}^t = -\frac{dN^S}{dt}, \tag{45}$$

where the coefficient $N^{\dot{q}} = \frac{d}{dt}N^q - \dot{q}\frac{dN^t}{dt}$ expresses the "prolongation" to $\dot{q}$ of the transformation on the (q,t) variable. Using the two first determining equations (20) (for $X = N^q, T = N^t, \phi = -N^S$), $N^{\dot{q}}$ can be rewritten as

$$N^{\dot{q}} = \frac{\partial \phi}{\partial q} - \dot{q}\frac{\partial X}{\partial q}.$$

In the stochastic case, i.e., along the random paths $t \mapsto z_t$, (36) tells us that the same hold true, replacing all derivatives $\frac{d}{dt}$ by their counterpart $\widetilde{D}$. The coefficient N^B in (36) is of the form

$$N^B = \frac{\partial \phi_N}{\partial q} - B\frac{1}{2}\frac{\partial X_N}{\partial t},$$

for ϕ_N and X_N solving the deformed determining equations (39) and can be, indeed, rewritten as:

$$N^B = \widetilde{D}X_N - \widetilde{B}\dot{T}_N$$

in conformity with the classical coefficient $N^{\dot{q}}$.

The invariance condition (36) implies the stochastic version of Noether Theorem (cf. [4]) which won't be needed here.

5. Examples, open problems and prospects

Let us consider the following ("free Boltz's") action functional for $[t,u] \subset I = [0, \omega^{-1}]$, and ω is a positive constant,

$$E_t \int_t^u \frac{1}{2}(\widetilde{D}z_\tau)^2 d\tau + E_t\widetilde{S}_u(z_u) \tag{46}$$

where $\widetilde{S}_u$ is of the form (31):

$$\begin{aligned} \widetilde{S}_u(q) &= -\hbar \ln \eta_u(q), \text{ for} \\ \eta_u(q) &= (1-\omega u)^{-1/2} \exp -\frac{\omega q^2}{2\hbar(1-\omega u)}. \end{aligned} \tag{47}$$

The diffusion extremal for this action (46) solves the SDE

$$dz_\tau = -\frac{\omega z_\tau}{1-\omega\tau}d\tau + \sqrt{\hbar}\, dw_\tau \tag{48}$$

or, in dynamical terms, equation (44) for $V=0$ with the boundary condition (47). This is checked by computing the drift $\widetilde{B}$ (41) using the solution $\eta(q,t)$ of the heat equation (32) for our given future boundary condition η_u, then substituting in the definition of $\widetilde{D}$.

It can be easier to use, instead of η_u, the following smooth initial condition for the heat equation (32):

$$\eta_0(q) = e^{-\frac{\omega}{2\hbar}q^2} \tag{49}$$

whose solution coincides, for $t \in [0, \omega^{-1}]$, with the one of the above final boundary problem.

A Theorem of Rosencrans tells us how to perturb the free heat Hamiltonian $H = H_0$ underlying the action (46) by, say, a quadratic potential V:

$$H = H_0 + \frac{\omega^2}{2}q^2 \tag{50}$$

in using only linear combinations of symmetry generators N of the free case (computed in [12]).

Then a solution of the perturbed heat equation, for the same initial condition as before, can be written explicitly in terms of the unperturbed solution. Here the two free generators needed to do that, among the 6 spanning the algebra $\mathcal{N}$ of §4 are

$$N_1 = \frac{\partial}{\partial t},$$
$$N_6 = 2qt\frac{\partial}{\partial q} + 2t^2\frac{\partial}{\partial t} + (\hbar t - q^2)\frac{\partial}{\partial S} + 2(q-tB)\frac{\partial}{\partial B} - (2qB + 4tE + \hbar)\frac{\partial}{\partial E}$$

and the resulting explicit relation between any solution of the free and perturbed (by $V = \frac{\omega^2}{2}q^2$) heat equations is given by

$$\eta^V(q,t) = (\cosh \omega t)^{-\frac{1}{2}} e^{\frac{\omega}{2\hbar}q^2 \tanh \omega t}\,\eta\left(\frac{q}{\cosh \omega t}, \frac{\tanh \omega t}{\omega}\right). \tag{51}$$

Using the definition (41), the corresponding relation between drift becomes

$$\widetilde{B}^V(q,t) = \omega q \tanh \omega t + \frac{1}{\cosh \omega t}\widetilde{B}\left(\frac{q}{\cosh \omega t}, \frac{\tanh \omega t}{\omega}\right). \tag{52}$$

This relation between stochastic derivatives can be integrated and provides the following one between the associated processes:

$$z^v(t) = \cosh \omega t \cdot z\left(\frac{\tanh \omega t}{\omega}\right). \tag{53}$$

By construction, any such diffusion $z^v(t)$ solves the dynamical equation (44) for $V(q) = \frac{\omega^2}{2}q^2$ i.e., is extremal point of an action functional (43) with this potential.

Coming back to our special boundary condition (49) (or (47)) characterizing the solution (48) of our free dynamical equation (z_τ is called the "Brownian Bridge") it is easy to check that the relation (50) reduces, then, to

$$\eta^V(q,t) = e^{-\frac{\omega}{2\hbar}q^2 + \frac{\omega}{2}t}. \tag{54}$$

Equivalently, using (52) for the drift $\widetilde{B}(q,t)$ of the Brownian Bridge given in the SDE (48), we obtain

$$\widetilde{B}^V(q,t) = -\omega q \tag{55}$$

$z^V(t)$ is the famous Ornstein-Uhlenbeck process, expressed here explicitly in terms of the Brownian Bridge $z(t)$ via the general relation (53), that we could call a "quadrature of $z(t)$" (cf. [11]) by analogy with classical dynamics. Indeed, this type of explicit relations expresses manifestly a kind of "stochastic integrability" for our underlying dynamical system. It says that, somehow, the two equations (44) for $V = 0$ and $V = \frac{\omega^2}{2}q^2$ are equivalent, or that there is a probabilistic version of symplectomorphism carrying one into the other. Since it follows from our construction that this relation is built on the one existing between two underlying HJB equations (27), we have here a simple example of integrability for this equation, resulting from our $\hbar$-deformation of the classical dynamical structure.

We believe that this strategy should allow us to construct a general Theorem of integrability for the n-dimensional Hamilton-Jacobi-Bellman, where all the ingredients of the classical mechanical proof are appropriately deformed. Such a proof would be useful for applications (given the versatility of (27)) but even more for its potential generalizations to other classes of processes, for example the one introduced in [15]. The explicit relations introduced between processes (like (53) for example) have a dynamical origin revealed only in second-order a.s equations like (44). Their systematization should introduce a number of new structures hidden behind the traditional theory of stochastic processes and their first order SDE.

Another interesting line of thought is to understand if it is possible (or not) to construct a probabilist interpretation not on the (q,t) submanifold as we did here but on the (q,B,t) submanifold of the jet bundle J^1, i.e., on a stochastic counterpart of the classical phase space. Of course, after "projection" on (q,t), the probabilistic interpretation looked for should reproduce the results given here.

Such a probabilistic theory on phase space would be very interesting. In the non rigorous "probabilistic" deformation created by R. Feynman, and known as the method of path integrals [16], it is striking to see that the status of the path integrals in phase space is much worse than the one in configuration space (which is supposed to have found its rigorous counterpart in Feynman-Kac formula. Cf. [4] for another viewpoint).

This allusion to quantum theory is, of course, not accidental. As an illustration, let us observe that if a.s second order equations like (44), resulting from characteristic equations (42), seem unusual in Stochastic Analysis, they are quite natural in a quantum perspective. Indeed, after expectation, namely

$$\frac{d}{dt}E[z_\tau] = E[\widetilde{B}], \quad \frac{d}{dt}E[\widetilde{B}] = E[\nabla V(z_\tau)],$$

they constitute probabilistic counterparts of Ehrenfest equations of motion for the position and momentum (Von Neumann [7]) observables of the system resulting from the quantization of Hamiltonian (25).

For the specific relation between what we did and quantum mechanics, the probabilistic content presented here is only one half of the story of "Euclidean Quantum Mechanics". The other half deals with the heat equation adjoint to (32) w.r.t. the time parameter (i.e., another filtration and its adapted backward SDE). These two adjoint PDE play the role of Schrödinger equation and its complex conjugate, and time reversible processes can be constructed from them, whose qualitative properties are very close to those of Feynman's informal "processes". Euclidean Quantum Mechanics is, in fact, our main source of explicit examples like the above-mentioned one, even if the geometrical ideas summarized here are relevant far beyond this initial motivation.

The analogy between our "imaginary time" framework and quantum mechanics is strong enough, for instance, to suggest new results in Hilbert space. This is what was found with the stochastic Noether Theorem alluded to at the end of §4 (cf. [17]). And this is also why we believe that the expected Theorem of integrability for HJB equation should help us to produce a reasonable definition of quantum integrability.

Acknowledgments

It is a pleasure to thank the local organizers for their new demonstration of the exceptional quality of Mediterranean hospitality.

References

1. P. Malliavin, "Stochastic Analysis", *Grund der Math. Wiss.* **313**, Springer (1997).
2. B. A. Dubrovin, I. M. Krichever and S. P. Novikov, "Integrable Systems I" in *Dynamical System IV (Symplectic Geometry and its Applications)*, eds. V. I. Arnol'd and S. P. Novikov, Springer-Verlag (1990).
3. J. L. Synge, "Classical Dynamics" in *Encyclopedia of Physics*, vol. **III/1**, ed. S. Flügge, Springer-Verlag, 1960.
4. K. L. Chung, J. C. Zambrini, *Introduction to Random Time and Quantum Randomness*, New Edition, World Scientific, 2003.
5. M. Giaquinta, S. Hildebrandt, "Calculus of variations II", *Grundl der Math. Wiss.* **311**, chapter 10, Springer, 1996.
6. B. K. Harrison, F. P. Estabrook, "Geometric approach to invariance groups and solution of partial differential systems", *J. Math. Phys.* **12**(4), 653 (1971).
7. J. Von Neumann, *Mathematical Foundation of Quantum Mechanics*, Princeton U. Press, Princeton, 1955.
8. A. Enciso, D. Peralta-Salas, "On the classical and quantum integrability of Hamiltonians without scattering states", to appear in *Theor. Math. Physics.*
9. P. Cartier, B. Julia, P. Moussa, P. Vanhove (eds.), *Frontiers in Number Theory, Physics and Geometry I*, Springer, 2006.
10. N. Ikeda, S. Watanabe, *Stochastic Differential Equations and Diffusion Processes* North Holland, 1981.
11. P. Lescot, J. C. Zambrini, "Probabilistic deformation of contact geometry, diffusion processes and their quadrature", to appear in *Proceedings Ascona 2005, Progress in Probability*, eds. R. Dalang, M. Dozzi and F. Russo, Birkhäuser.
12. P. Lescot, J. C. Zambrini, "Isovectors for the Hamilton-Jacobi-Bellman equation, formal stochastic differential and first integrals in Euclidean quantum mechanics", in *Progress in Probability*, vol. 58, eds. R. Dalang, M. Dozzi and F. Russo, Birkhäuser, 2004, p. 187.
13. A. B. Cruzeiro, J. C. Zambrini, "Malliavin calculus and Euclidean quantum mechanics. I Functional Calculus", *J. Funct. Anal.* **96**, 62 (1991).
14. D. Lovelock, H. Rund, "Tensors, Differential Forms and Variational Principles", Dover Publications, New York, 1989.
15. N. Privault, J.C. Zambrini, "Markovian bridges and reversible diffusions processes with jumps", *Ann. Inst. H. Poincaré* **PR40**, 599 (2004).
16. R. P. Feynman A. R. Hibbs, "Quantum Mechanics and Path Integrals", McGraw-Hill, New York, 1965.
17. S. Albeverio. J. Rezende and J.C . Zambrini, "Probability and quantum symmetries II — The theorem of Noether in quantum mechanics", to appear in *J. Math. Physics* (2006).

JEAN-CLAUDE ZAMBRINI
Grupo de Física-Matemática da Universidade de Lisboa,
Av. Prof. Gama Pinto, 2, 1649-003 Lisboa, Portugal
zambrini@cii.fc.ul.pt

List of participants

Luigi Accardi · *Roma 2*

Ahmed S. Al-Rawashdeh · *Jordan University of Science and Technology*

Hélène Airault · *Picardie*

Abdul Rahman Al-Hussein · *Al Quassam, Saudi Arabia*

Ahmed Al Salam · *Yarmouk*

Wided Ayed · *Tunis*

Abdessattar Barhoumi · *Tunis*

Sonia Belkadhi-Chaari · *Bizerte*

Mohamed Ben Chrouda · *Monastir*

Anis Ben Ghorbal · *Roma 2 / Tunis*

Zied Ben Salah · *Tunis*

Madi Belcacem · *Guelma, Algerie*

Philippe Blanchard · *Bielefeld*

Wissem Boughamoura · *Tunis*

Radhouan Boukhris · *Tunis*

Mounir Bezzarga · *Tunis*

Refaat Chaabouni · *ENIT*

Fernanda Cipriano · *Lisbon*

Ana Bela Cruzeiro · *Lisbon*

Houcine Chebli · *Tunis*

Sidi Hamidou Djha · *Tunis*

Ahcene Djoudi · *Annaba*

Vito Crismale · *Bari*

Mohamed El Oued · *Monastir*

Nejib Ghanmi · *IPEIN*

Soumaya Gheryani · *Tunis*

Diogo Gomes · *Lisbon*

Hanno Gottschalk · *Bonn*

Martin Grothaus · *Kaiserslautern*

Skander Hachicha · *Tunis*

Rached Hachaichi · *Tunis*

Ezeddine Haouala · *Tunis*

Takeyuki Hida · *Meijo*

Sameh Horrigue · *Monastir*

Ma'rton Ispa'ny · *University of Debrecen, Hungary*

Raoudha Jenane-Gannoun · *Tunis*

Riadh Kaddachi · *Tunis / Mannheim*

Mounir Khlifi · *Tunis*

Paul Krée · *Paris 6*

Yuh-Jia Lee · *University of Kaohsiung*

Paul Lescot · *Saint-Quentin, France*

Paul Malliavin · *Paris 6*

Penka Mayster · *Tunis*

Rui Vilela Mendes · *Lisbon*

Giulia Di Nunno · *Oslo*

Nobuaki Obata · *Tohoku*

Bernt Oksendal · *Oslo*

Maria J. Oliveira · *Lisbon*

Habib Ouerdiane · *Tunis*

Andrea Posilicano · *Italy*

Nicolas Privault · *La Rochelle*

Hafedh Reguigui · *Tunis*

Anis Riahi · *Tunis*

Barbara Rüdiger · *Koblenz*

Luís Silva · *Madeira*

Nejib Smaoui · *Kuwait University*

Boubaker Smii · *Tunis / Bonn*

Fethi Soltani · *Tunis*

Lakhdar Tannech Rachdi · *Tunis*

Alex Uglanov · *Yaroslavl, Russia*

Wilhelm von Waldenfels · *Heidelberg*

Jean-Claude Zambrini · *Lisbon*